T0300257

Nanotechnology

Business Applications and Commercialization

Nano and Energy

Series Editor: Sohail Anwar

Computational Nanotechnology: Modeling and Applications with MATLAB®
Sarhan M. Musa

Nanotechnology: Business Applications and Commercialization
Sherron Sparks

Nanotechnology

Business Applications and Commercialization

Sherron Sparks, PhD

CRC Press
Taylor & Francis Group
Boca Raton London New York

CRC Press is an imprint of the
Taylor & Francis Group, an **informa** business

CRC Press
Taylor & Francis Group
6000 Broken Sound Parkway NW, Suite 300
Boca Raton, FL 33487-2742

© 2012 by Taylor & Francis Group, LLC
CRC Press is an imprint of Taylor & Francis Group, an Informa business

No claim to original U.S. Government works

Version Date: 20120224

International Standard Book Number: 978-1-4398-4521-9 (Hardback)

Visit the Taylor & Francis Web site at
http://www.taylorandfrancis.com

and the CRC Press Web site at
http://www.crcpress.com

Dedication

I learned about nanotechnology in 2004 in a darkened room in Queens Hospital in Honolulu. My brother Ned lay critical, his body ravaged by the AIDS virus, and told me of this wondrous new discovery that would someday save lives. In the wee hours of the night, we (Ned; my mom, Verna Mae; and sister, Rose Ann) talked about life, learning, family, friends, and nanotechnology. It is where I became enamored with the potential for this magnificent science and what it could do for this world and humanity. That night opened a whole new world for me and thus this book.

To Ned, Mom, and Rose Ann for the Hawaiian discovery of Buckminster Fuller and nanotechnology.

Besa

Contents

Disclaimer

This book is intended for general information purposes only. Although all care has been taken to ensure the accuracy of the information, Dr. Sherron Sparks does not accept any responsibility for any errors and omissions. Dr. Sparks does not accept any responsibility or liability for any loss to any person acting or refraining from action as the result of, but not limited to, any statement, fact, figure, expression of opinion, or belief contained in this book.

Preface

I have always lived my life with eyes open in wonderment. I count myself blessed to have the inquisitive nature of a child. The world is full of wonderful new discoveries if we just open our eyes, hearts and minds to this vast universe that we are privileged to journey for a period. I love learning. I love research. I love discovering. Wish that all humanity could go on this quest for discovery and leave the wars, anger, clandestine politics and distrust behind and give their lives to making a positive difference for mankind, rather than destruction.

I look at nanotechnology as the means some will take to pursue miraculous discoveries that can enhance our lives, heal our cancers, grow the food, destroy harmful viruses and bacteria, freshen our water, eradicate our disease, plump our wrinkles, and in general, help us to live in a new world via the powers of the electron microscope. In effect, it is a chance to view through a new lens, the life we are privy to in this 21st century. No time has ever been so exciting, moved so quickly, nor has any other time embodied the world in the same way as nanotechnology has the power to do. It is time to take off our blinders, shed our fears and step into the power of this gift we have been given. It is time to harness and standardize the growth to protect us all. It is time to "go boldly where no man has gone before…"

What we do with this gift that is before us falls back in the lap of humanity.

As you will see in the following pages of this book, the wonderment known as nanotechnology has the capability to touch virtually every aspect of our lives. New discoveries have touched our lives in products that we have been unaware we were using. The secrets of nanotechnology genius were leaked to the world over the past few years by modern-day horror writers, who challenged us to be wary of the power that can be unleashed. Brilliant writers have researched the science through the common man's eyes and shared how zealots can potentially abuse it. Noted scientists have published papers on the wonders, and yes, the dangers, as they spend their lives poring over the microscopes watching miracles unfold before their eyes. Governments have synthesized the data and realize the economic potential that could come from being a leader in the realm. Investors, venture capitalists, and financiers are becoming interested in the return on investment (ROI) that could create wealth of vast proportions. The military has given it over to major universities to come up with new ways to protect our men and women fighting abroad, as well as to make weapons of war. Educational systems are beginning to offer primary classes and terminal degrees. Businesses are taking notice and starting to question their involvement. And surprisingly, a large part of the world has never heard of it.

I wrote this book because the time has come. For the most part, the miracles of nanomaterial technologies still lie in the catacombs of the great universities awaiting commercial birth; a way to propel it through the "valley of death" that many technological discoveries must traverse seeking fruition. I have strived to create a nanotechnology commercialization model that will enhance the partnership among universities, venture capitalists and business to eliminate this commercialization gap, thereby providing a structure for industry opportunity and growth of technology applications.

There is great work to be done. It is time to focus—to put forth our efforts to channel the development of this amazing technology via legitimate business channels, standardizing the development throughout the world, so that it can emerge as a safe, useful technology that limits the hazards of shoddy development. This delicate technology must not be suppressed out of fear only to land in the hands of zealots with unsavory power ideals; rather it should be unleashed to mankind through commercialism and the businesses community in applications that will benefit the world. We must create legitimate structures for growth and business opportunities, rather than to leave this power to a few. By bringing it to the commercialism and business realm via standardization, individuals can be a part of it, making it a technology that can be used by the masses to create jobs, stimulate the world economy and better mankind.

As a doctor of industrial/organizational psychology, I look upon not only this wondrous science for the miracle that it is, but also to how it fits the scheme of survival in our world. As an I/O psychologist, I study the workplace, applying psychological principles to issues of critical relevance to the business community—in this case organizational development with regard to a new technology breakthrough that could usher in the next industrial age, thereby stimulating our workforce and economy.

This book was designed for students, investors, scientists, politicians, universities and the everyday individual to open a dialogue about our future. It is a clarion call—an urgent call to action to move the technology from the laboratories to the business world, rather than suppress the genius of our scientists. So many great discoveries have died in laboratories, or were sacrificed for unnatural gains. I am enthralled by Einstein. I would love to have a discourse with him today on how his ideas could have been used to further humanity and our planet, and not be used as political power by rogue governments.

This book will take you through the journey of nanotechnology to commercialization, looking at the various industries it affects, ease of entry options, intellectual property concerns, emerging ethics, its dangers, standardization needs, investor resources, government support, various business applications, supporting organizations, and equally important, the social ramifications.

About the Author

Dr. Sherron Sparks earned her master's degree in organizational leadership and a PhD in industrial–organizational (I/O) psychology. She is president and CEO/CLO of Sparks Business Institute, an end-source training and executive/leadership coaching entity for high-achieving executives in succession for CEO/COO positions. In addition to her business endeavors, she is also an instructor at The Pennsylvania State University, teaching business, psychology, and labor and industry classes.

Dr. Sparks was introduced to the field of technology while serving as general manager and COO of an international technical business. It was while in that position that her interest in the practical applications for nanotechnology was born.

Always at the forefront of industrial and organizational trends, Dr. Sparks' ability to foresee cutting-edge business and industry trends has led to international recognition as a business leader and as an instructor. Dr. Sparks is a published writer who saw the importance of the emerging and exciting field of nanotechnology, and how it is already impacting and will impact the marketplace. In 2002, her firm was selected as the only woman-run technical company, in a group of 15 companies, to accompany the governor of Pennsylvania to Australia to study new technology trends.

Acknowledgments

Building this wealth of knowledge, I was privy to the great minds of multiple people involved in the tremendous growth and support of nanotechnology. I want to acknowledge and thank them for their contributions, support, friendship, and encouragement as I completed this work: Dr. K. Eric Drexler, Dr. Stephen Fonash, Dr Alan Brown, Dr. Richard Doyle, Kartik Puttaiah, Mark Charleston, Joe Rubin, Dr. Akihisa Inoue, Yoshinori Yamamoto, Judith Lightfeather, Liam Glessner, Betsy and Mylee Glessner, Kenneth D. Miller, II, Warren Miller, Charles Miller, Michael Hain, and Elaine Long.

The Pennsylvania State University, Rice University, Smalley Institute, the U.S. Department of Commerce, Pennsylvania NanoMaterials Commercialization Center, Nanogate, U.S. Patent and Trademark Office (USPTO), Invn-Tree, NACK Center, Funding Post, Tohoku University, The NanoTechnology Group, Inc., Glessner Photography, Sparks Business Institute, and WHVL TV in State College, Pennsylvania.

I want to give a special acknowledgment for the tremendous support in this book from Dr. Sohail Anwar, the editor of the Nanotechnology Series and the editorial staff of CRC Press—Nora Konopka, Robin Lloyd-Starkes, and Jessica Vakili— as well as the rest of the production team at CRC Press.

To the countless students, friends and family who took the time to listen and believe in me—thank you all so very much.

Learning is a treasure that will follow its owner everywhere.

—Chinese proverb

1

Introduction

"I sit before you today with very little hair on my head. It fell out a few weeks ago as a result of the chemotherapy I've been undergoing. Twenty years ago, without even this crude chemotherapy, I would already be dead. But 20 years from now, nanoscale missiles will target cancer cells in the human body and leave everything else blissfully alone. I may not live to see it. But I am confident it will happen." Richard Smalley, a 62-year-old Nobel Prize–winning chemist, was a nanotech pioneer who spoke these words on June 22, 1999. He died of non-Hodgkin's lymphoma on October 28, 2005.

A little more than 5 years after his death and the process has begun. We are currently in the midst of a massive new industrial revolution—on the nanoscale—that has the ability to change the world as we know it. From his seed of an idea, to a worldwide effort to commercialize this disruptive technology, one can only wonder what the next 5 years will bring with the exponential changes we are now seeing in this field.

The discovery happened when Dr. Smalley and fellow Nobel laureates Robert Curl and Harold Kroto came together to explore Kroto's interest in red giant stars that were rich in carbon. In trying to recreate the environment in a lab, Curl suggested they contact Smalley, who had built an apparatus that could evaporate and analyze almost any material with a laser beam. The three joined together for a week in Houston in 1985 with younger coworkers, James R. Heath and Sean O'Brien, and starting from graphite, produced a cluster of 60–70 stable carbon atoms.

Dr. Smalley and his colleagues made the completely unexpected discovery that the element carbon also existed in the form of stable spheres or carbon balls that they called fullerenes. These carbon balls were formed when graphite was evaporated in an inert atmosphere, developing around the balls, creating a new carbon chemistry that made it possible to enclose metals and noble gasses in them to form new superconducting materials and create new organic compounds and polymeric matter.

Dr. Richard E. Smalley was the 1996 recipient of the Nobel Prize for Chemistry for the discovery of fullerenes. He served as chairman of the Rice Quantum Institute, was the head of the Center for Nanoscale Science and Technology at Rice University, and chairman of the board for Carbon Nanotechnologies, Inc. Dr. Smalley was also on the scientific advisory board of CSIXTY, Inc., director of the Rice Center for Nanoscale Science and Technology (CNST), and director of the Carbon Nanotechnology Laboratory.

Smalley's C_{60} carbon cluster resembled architect Buckminster Fuller's spherically designed building for the 1967 World Exhibition in Montreal. Buckminster Fuller had used hexagons and multiple pentagons to create "curved surfaces." Smalley's group assumed the C_{60} consisted of 12 pentagons and 20 hexagons with carbon atoms at each corner—the same form as a European football. The group called the carbon ball C_{60} buckminsterfullerene. Using colloquial English, the carbon balls became known as buckyballs.

This 1985 discovery has grown, almost silently, within the labs and brains of those studying the significant and, at times, almost unbelievable uses. Today there is a nano buzz in the air. The nano revolution is beginning to gain attention—it is no longer the hidden genius of scientists, the brainchild of horror writers scripting grey goo, or a lofty technology that only the intelligentsia knew about. It is coming to the populace in products manufactured for everyday life; products that can heal, save lives, be more durable, last longer, save our armed forces in the field, etc. It is gaining the attention of the business world with investors wanting to ride the next industrial revolution of technology.

Nanotechnology has long been seen as the last industrial wave of the future, and governments and universities have been pouring billions into its development. Now it is time for industries, investors, and businesses to follow suit, to partner with universities to make this wondrous technology, which can literally change the world molecule by molecule, a commercial reality (Figure 1.1).

Anyone who has ever owned a business or has the entrepreneurial drive to research, the first thing one experiences is elation—then feeling overwhelmed. There are mountains of information to sift through and form into a workable commercial business plan that supports the idea. One needs facts. Plenty of them. Enough information should be gathered so that one knows whether shirts will be lost or dreams fulfilled as a result of this venture. How many times have you said aloud, "I wish I had a book that had all this information in it for me so I didn't have to spend months researching it for myself."

That is what this book on the commercialism of nanotechnology offers—information that you can put your finger on, a guide you can go to find information to become a part of this wondrous new technology, called nanotechnology, that is rapidly changing the world. This book is a guide for scientists, investors, university students, the public, and those interested in the future of nanotechnology. It does not hold all the answers, technology moves too fast. But, it is a knowledge framework that you can take to the bank. We, as business people who want to ride this massive wave, need to have a clear understanding of the technology, what it is capable of, where our particular skills and talents fit into its growth, and then identify the direction and focus of nanotechnology, as it pertains to our interests. We need to identify the barriers that constrain the commercialism of this disruptive technology.

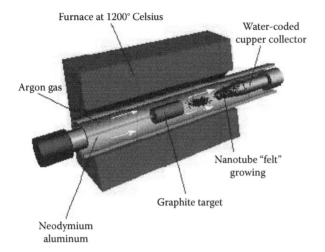

Laser Ablation or Pulsed Laser Vaporization (PLV)

A laser is aimed at a block of graphite, vaporizing the graphite.
Contact with a cooled cooper collector causes the carbon atoms
to be deposited in the form of nanotubes.

© *American Scientist* 1997

The nanotube "felt" can then be harvested, thus creating a nanotube

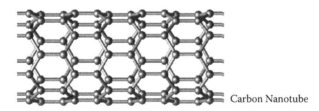

FIGURE 1.1
Pulsed laser vaporization growing nanotube *felt*. Nanotube "felt" growing along tip of collector; neodymium–yttrium–aluminum–garnert laser.

So, what is nanotechnology and how do you commercialize something you can't even see with the human eye? Read on!

Microscopes, both optical and atomic force, like the one shown in Figure 1.2, of an ant's retina, are used for biolife projects and nanotechnology to view life at the nanoscale. Nanotechnology is the manipulation (or self-assembly) of individual atoms into structures that can create materials and devices with new or vastly different properties. It works in two ways. Nanotechnology can work from the top down, reducing the size of the smallest structures to the nanoscale, such as the application of photonics in nanoengineering and nanoelectronics. Nanotechnology is also known as molecular manufacturing. It can also work bottom up, manipulating individual atoms and

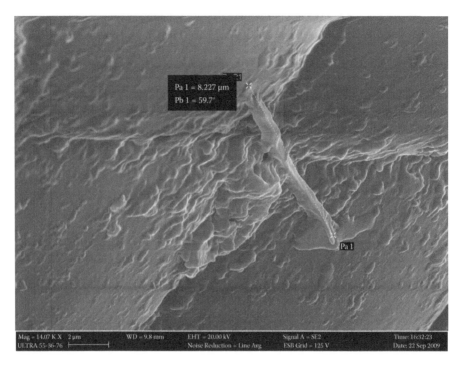

FIGURE 1.2
An ant retina, captured with a scanning electron microscope (SEM) at The Pennsylvania State University, showing the size comparison at the nanolevel. (Photo courtesy of The Pennsylvania State University.)

molecules into nanostructures similar to chemistry or biology structures. It has the ability to touch every facet of our lives.

What makes this molecular manufacturing so unique is that although it works on such a small scale, its impact as a disruptive technology that can touch so many products is tremendous. Working at the nanolevel, there is a high surface-to-volume ratio that has most of the atoms on the surface of the elements on which scientists are working. This all accounts for a unique surface environment for the atoms, as the surface forces dominate bulk forces (e.g., gravity). At this atomic level, the size of the properties correspond more closely to macromolecules and basic biological structures, one of the factors that allows for chemical bonding configurations. This is also a size range in which molecules can self-assemble.

As mentioned earlier, nanotechnology is considered a disruptive technology. A disruptive technology changes or disrupts the existing product with innovations to improve it in ways that were never considered, such as designing a new way to make a product more efficient or significantly lowering the price. It disrupts the existing market. Words like *evolutionary* or *transformational* explain it best. It is a new way of doing things that disrupts the

FIGURE 1.3
An individual can visit the local drug store and peruse through several products that use nanotechnology. L'Oreal RevitaLift Double Lifting has an intense retightening gel that uses pro-tensium plus nanosomes of proretinol A.

current market. One example is the invention of the automobile. Advances in technology and design have made cars safer and more energy efficient. The innovations to autos have changed the market almost yearly to give us the automobiles we have today.

Thus nanotechnology is not a new product, but a potential enhancement to nearly all existing products. Another example is the current cosmetic industry. In our health- and beauty-conscious society, both men and women are looking for the product that will ease the wrinkles without surgery or shots (Figure 1.3).

Nanosome technology utilizes the most effective tool for bidirectional delivery of both water- and oil-soluble ingredients in the skin. Lancôme's® Hydra Zen® and RevitaLift® contain miniscule nanocapsules of chemicals that cosmetics have used for years. Other skin creams, sunscreens, etc., use particles of silica and nanoparticles of zinc oxide. Zelens Day Cream® uses buckyballs (buckyballs are carbon atoms a billionth of a meter wide) in its formula to prevent premature aging of the skin by acting as an antioxidant. The chemicals are put into the buckyballs, and because they are so miniscule, they carry the wonderworking material into the fine lines and wrinkles rather than just glossing over them.

But wait—nanotechnology? Isn't a technology something? The common person is used to the word *technology* being associated with computers, chips, and devices. Sound confusing that something that you thought of as a normal technology is in our makeup?

Drilling it down to something imaginable, a nanometer is one billionth of a meter—1,000 times smaller than the microscale that was traditionally associated with the electronics industry. The term *nanotechnology* refers to the engineering measurement of the nanoscale materials and devices. To give an idea of the size of a nanometer (nm), here are some examples:

- Typical red blood cell, 7,000 nm in width, 2,000 nm in height
- Common cold virus, 25 nm
- Width of DNA molecule, 2 nm
- Silicon atom, 0.2 nm

What makes nanotechnology so amazing as a new technology is the fact that it is not a single product or group of products, but that at the atomic level, it represents the entire scientific field, enabling creation of multiple applications in multiple industries.

Numerous materials can be engineered into nanoparticles, using small clusters of atoms of gold, silver, iron, zinc, silica, titanium dioxide, etc. Carbon is another material used and can be made into hollow balls or tubes of atoms known as *fullerenes*. Breaking these materials down to the nanoparticle stages brings about new properties. For example, silver at this nanoparticle stage is very effective at killing microbes and keeping medical and food appliances hygienic; iron, on the other hand, has been found to be very effective at removing pollution from contaminated land. Scientists are discovering more and more uses for the technology around the world and are looking to bring many of these products to market. The following are just a few of the products currently developed or under development around the world:

- Diagnostics for doctors' offices in the form of point of care sample analysis that will speed a doctor's ability to diagnose and treat medical conditions in a shorter timeframe, thereby saving lives.

- *In vitro* diagnostics with ultrasensitive nanomarkers that can detect cancer and other diseases.

- Nanotech soldiers, outfitted with lightweight, flexible body armor that stops bullets, splints and casts, gives CPR, administers life-sustaining medications, and uses the body's energy, movement, etc., to create power supplies for the body gear.

- A villager in Africa scoops a handful of fresh cow manure and puts it into a hanging "filter" tied to a tree in the village square and waits for his badly needed drink of fresh water to drip out of the bottom of the filter.

- Replace half the lighting in the nation with new energy-efficient LED bulbs; the annual savings would be equivalent to the annual output of 50 nuclear reactors.

- Dental applications that include filler systems consisting of nano-clusters and nanoparticles that are stronger and smoother while providing color-matching capabilities because the nanoparticles are smaller than a wavelength of light, making them readily transparent.

- Diagnostic biochips that detect malignant or premalignant lesions in less than 15 minutes from a swab taken from inside the patient's cheek; normal blood tests would take days to return from the lab.

- Nanobiochips that can analyze biomarkers for multiple diseases, such as HIV, trauma injuries, and heart disease.

- Nanorobots in your bloodstream used to find, recognize, and destroy the particular microbial strain.

- "Grass-powered" solar energy on roof tops.
- "Injectable medical bombs" that can trace out a cancer and destroy it within, taking the place of invasive and painful caustic treatments.
- Treated fabrics, giving you socks that don't have foot odor after wearing.

Could this amazing science be the next industrial era, the financial bubble that will carry multiple nations to prosperity? Is it the magical cure? Or, is it a dark fantasy that has come to life?

"By 2015, industry analysts estimate that consumer spending on nanotech-enabled products could reach $12.5 trillion annually on upgraded, everyday products, super-electronic communications, and life-saving medical devices," said Mitch Horowitz, director of strategy at the Battelle Technology Partnership Practice.

Nanochip, Inc., located in Fremont, California, was founded in 1996 and was a very small company until 2003. Nanochip has 8 granted U.S. patents and 34 more patent applications. Today, about 50 people worldwide are working together to bring our microelectromechanical system MEMS storage chips into production in 2010. The company is developing a new class of ultra-high-capacity storage chips based on MEMS technology. MEMS - Microelectromechanical systems (MEMS) is the technology of very small mechanical devices that are driven by electricity; at the nano-scale it is nano-electromechanical systems (NEMS). MEMS are also referred to as micro-systems technology or MST in Europe and micromachines in Japan. This technology will enable the storage of tens of gigabytes of data per chip, the equivalent of many high-definition feature-length videos, at a substantially lower cost than today's flash drive memory sticks. Nanochip uses a three-waver stacked MEM technology that produces 32 Gbytes per die, eventually gaining to 4 Tbytes per die by 2017. "Hopefully by then you'll have 10 to 20 terabytes in your cell phone," said CEO Gordon Knight (Figures 1.4 and 1.5).

With an estimated $11.7 billion in sales in 2009 (originally predicted to be $9.4 billion), BCC research predicts a compound annual product growth of 11% through 2015, and looking at a 5-year goal of $26 billion in sales. The National Science Foundation estimates that by 2015, the United States will command about 40% of the trillion dollar worldwide market for nanotech products and services.

To many this looks like the industrial boom that will pull the world from the economic crisis in which we find ourselves.

And yet, there is a segment of the population, both lay and those in the scientific community, that sees the nanotechnology explosion as a "cure worse than the disease" (Freitas, 2009), citing self-replicating nanofactories, zealots looking for revenge, and inhalation of dangerous particles that some believe are worse than asbestos, as just a few of the fears.

With the dangers looming, why write a book on the commercialization of this space age fantasy?

FIGURE 1.4
A child born in 2010 will be submerged in the technology by 2015. Nanotechnology will be Liam's way of life. While we may struggle with the concept of a gigabyte, this generation will understand and live in a world where the reality is multiple Tbytes. (Courtesy of Glessner Photography.)

FIGURE 1.5
Richard Smalley, 1996 Nobel laureate in chemistry, Rice University.

The answer is simple. There is no way to stop the growth of nanotechnology—it is in far too many hands, and has the proven capacity to be a life-changing power for good. Fear and banning the use of the technology would push the research underground, putting it into the hands of "unaccountable scientists, rogue regimes and often under inadequate safety precautions" (Freitas, 2009).

This book is written for business professionals as well as the scientists and physicists, venture capitalists, etc., combining their disciplines, giving them insight into each other's strengths, as each strives to bring nanotechnology to the world in a safe, profitable venue. It is a guide to harnessing the intelligence of this fascinating technology through the application of strong business principles, driving the standards and development within the right circles, and taking the knowledge to the commercial level through proven business applications for the betterment of mankind, science, and businesses throughout the world.

We have only to look back at the technical boom of the dot-coms to learn a great lesson in the management of technology. Creative genius will not sustain a business. One must apply business practices and principles to harness the power of the creativity to become competitive in the marketplace. As with most creatives, the project becomes your baby, which you do not want to let out of your sight or control, but often the tight control can lead to poverty and abuse of what could be great. I liken the commercialism of any new business to the African concept of raising a child: "it takes a village to raise a child." And it takes a village of support to raise a creative idea into a business. From angel investors and bankers to employee management and solid business principles, your commercialism journey with your nano idea will now begin a solid growth into the world of business and corporations.

There is often spoken of an almost unconquerable divide between universities genius and the commercialism of the product—the valley of death. This almost insurmountable chasm to date was the reason for the dot-com explosion of young gifted men and women jumping into the river that runs between and giving their lives and dreams to make good on their ideas. Some made it—Bill Gates. Others made a small fortune but lost it because they did not have the business acumen to carry their dreams the entire way to the side of commercial success. Taxes, employee problems, logistic problems, competition, etc., overwhelmed them in waves and they drowned when they were so close.

It can be a precarious river to cross, but having a good map is key. I personally stepped into the realm of the technical boom and had to harness the discordant activities, bringing business discipline and accounting measures for value-added services that became the cash cow, building teams, hiring a stable of technical employees with vision as well as technical skills, etc., taking a feast-or-famine company struggling for rent, in a dingy office, to a respected and known company that began to go global before its sale. I, in no way, did it myself; I just created the framework for success and let the players do their jobs. This is the secret to commercialism—getting into the right village (environment) and letting the village that you trust help to raise your idea to commercial success.

This is where savvy business professionals can step in. Learning through the dot-com debacle, the business world understands that the creative genius of the new technology needs to partner with a solid business foundation to bring the dream to fruition. Business practices and principles are critical to sustain any business. Most scientists do not have this set of practical business skills needed to garner the funding, set up a viable enterprise, and hire the right people to make their vision a reality, just as most business persons do not have the technical savvy needed to create a product. Thus the marriage of business and science is needed to cross the chasm of issues called the valley of death to commercialism.

Taking a look at the reasons behind the valley of death is imperative. Before taking corrective action, figuring out the details of the problems of failed companies is helpful. Why have these nanotechnology companies failed at the same particular valley of death stage? The U.S. federal government spends billions of dollars on different types of basic nanotechnology research. Venture capitalists are willing to invest billions more on nanotechnology products once a credible business plan and team have been assembled. However, in the challenging period between these two stages, there is a significant gap in financial capital available to nanotech firms.

With a few exceptions, such as some Small Business Innovation Research (SBIR) and Defense Advanced Research Projects Agency (DARPA) grants, the federal government has so far been unwilling to finance research efforts within this gap, while venture capitalists are reluctant to take on the substantial risk that the precommercialization investment may never become a marketable product. This gap period in capital financing, commonly referred to as the valley of death, is where good lab discoveries go to die because they lack the funding necessary to become a commercial product. Nanotech innovations are particularly at risk for succumbing to the valley of death. Burned from the dot-com bust, venture capitalists are all the more unwilling to finance technologies that are not close enough to being a saleable product, especially "platform technologies" that have yet to prove their effectiveness on the market.

Jacob Heller and Christine Peterson, writing the "Valley of Death in Nanotechnology Investing" for the Foresight Institute, feel there is reason for concern regarding a valley of death in nanotech-based industries.

> Nanotech venture capital financing has accumulated in a concentrated set of well-developed firms. For example, during the first quarter of 2005, all venture capital deals in nanotechnology went to only four large, well-known firms. Venture capital investment is so low that all VC nanotech investment from 1998-2004 is approximately equal to the amount that the government spent on nanotechnology in 2004 alone. The Valley of Death is a serious concern, and is seen by many industry leaders and members of Congress as the main roadblock preventing nanotechnology industries from reaching maturity. If good ideas do not survive through the valley and come out as commercial products, the initial research results lie unused and monies spent wasted. End products are the inventions that actually benefit humanity and drive the economy, and they are seriously delayed. Members of Congress on both sides of

the aisle have come up with myriad propositions meant to deal with the valley of death. Representative Mike Honda (D-San Jose) introduced the Nanomanufacturing Investment Act of 2005 (HR 1491). The act would make available $750 million in government financing and investments for nanomanufacturing projects (and explicitly *not* basic research, which the government already finances through the National Nanotechnology Initiative), contingent on private investments of $250 million. Private investors would get a greater return on their investments to persuade them to make these relatively risky investments. Funding applications would be peer reviewed and approved by an advisory board. The program would continue until all $1 billion is invested.

In testimony to Congress, Sean Murdock, the Executive Director of the NanoBusiness Alliance, outlined other possible solutions to bridge the valley of death. His solutions include increasing support for infrastructure and nanotech user facilities, which would provide necessary machinery and tools to nanotech developers and therefore decrease the capital requirements of nanotech startups. Murdock also suggested linking basic research with specific goals, increasing federal funding for SBIR and the Advanced Technology Program, and increasing overall investment in nanotechnology research. While the above represent mainstream suggestions for solving the valley of death problem, many observers are skeptical of additional government involvement in technology commercialization, advocating a more free-market approach. Policy options, which reduce the burdens on startup companies, in general should also be explored, from reform of the Sarbanes-Oxley Act at the federal level, to the abolishment of California's unusual tax on the purchase of manufacturing equipment. A wide variety of policy options, some free-market and others interventionist in flavor, have been outlined by the Blue Ribbon Task Force on Nanotechnology. Regardless of which strategy governments choose to address the Valley of Death, action is desirable to speed nanotech products to market. Some creative programs could possibly help nanotech entrepreneurs in their efforts to bring the benefits of nanotech advances to everyone in an affordable way. (Heller and Peterson)

This era of turbulence is truly one of the most exciting times in our lives. While economic woes, wars, and devastation are shaking our world to its core, we stand in awe at the timely birth of the creation of nanotechnology innovations to what might be the answer to many of our problems—cures for AIDS and cancer, a new industrial age that can boost the global economic fears, and an influx of jobs.

The burning question is: How do we harness and commercialize this new technology that is straining to take on a life of its own? In 2007, the University of Illinois at Springfield prepared a report for the U.S. Department of Commerce entitled "Barriers to Nanotechnology Commercialization." Authored by Ronald D. McNeil, PhD, Jung Lowe, JD, Ted Mastroianni, MPA, Joseph Cronin, EdD, and Dyanne Ferk, PhD, the report gives insight into the strengths and weaknesses that the United States currently experiences in the commercialization of nanotechnology.

The purpose of the Department of Commerce study by the University of Illinois at Springfield was to identify barriers that hinder the commercialization of nanotechnology. The data were created to give policy makers more concrete information to make decisions regarding the progression of nanotechnology for U.S. companies and American workers to best capitalize on the existing R&D available. The information was collected from key nanotechnology stakeholders, scientists, university centers, venture capitalists, private companies, researchers, and members of the National Nanotechnology Initiatives, etc.

This study, which serves as a backbone for U.S. policy makers, will be referred to throughout this book as we take scientific discovery of nanotechnology to commercial products. Other key information and studies will be referred to as the experts in the industry give reports on this platform technology.

The following are the key barriers to the commercialization of nanotechnology identified in the U.S. Department of Commerce report that act as a guide how to navigate this new nanoworld:

CASE STUDY: KEY BARRIERS TO NANOTECHNOLIOGY COMMERCIALIZATION

Relevant Barriers to Nanotechnology Commercialization
U.S. Department of Commerce—2007 Report

1. Time between research and commercialization is estimated to be 3 to 10 years. Venture capitalists and other sources of funding find this time factor to be a detriment.
2. The so-called valley of death is the often fatal interlude between scientific results of the researcher and initial funding for prototyping and commercialization. The scientists may publish results and not be interested in commercialization. As often happens, where there is interest or not in commercialization, the common comment is that for every dollar invested into basic research, which is critical to U.S. competitive strength, almost $100 is required for a competitive product to be produced. The commercialization of nanotechnology scientific investment has little relationship to the hi-tech dot.com software commercialization paradigm. This is a serious gap between research and commercialization that must be addressed by government agencies and venture capitalists.

3. Lack of proper infrastructure (labs, equipment, measuring devices, etc.) hinders the growth of small business and researchers. The infrastructure needed is very expensive. Furthermore, equipment becomes quickly outdated due to the major advances in technology.
4. Lack of usage of federal and university laboratories and equipment hurts small businesses that can't afford this infrastructure.
5. Many of the employees or scientists are foreign nationals. They are not allowed access to federal labs in most cases.
6. Small businesses do not have the capacity to produce products at a large scale.
7. There is a lack of a coherent policy on tech transfer from universities to startup businesses.
8. Audit control from federal government is a hindrance to small companies. It is very expensive to slow down work to comply with several federal agencies that conduct audits. There needs to be a centralized system.
9. Patent office takes up to 36 months to respond to applications registered.
10. Potential barriers may include the lack of trained scientists, engineers, technicians, and researchers in this country. There is no federal policy addressing the deficit in scientific training at all levels of our educational institutions and in improving the workforce with better and improved technical skills.
11. The current tax policy does not assist research and development. There are not enough sufficient tax credits for funding groups.
12. FDA and Patent offices do not have enough qualified staff to assess nanotechnology products.
13. The development of nano tools must increase and be more available to universities and startup businesses.
14. SBIR encourages research and not commercialization. It does not support small companies.
15. Applied research needs to be encouraged more in universities and federal labs.
16. The public perception that nanotechnology products are unsafe must be challenged to insure the public fully understands its potential.
17. Lack of standards and measurements are hindering advancements in nanotechnology.
18. The reduction of research and development funding has been hindering advancement in research.

19. Current immigration policy is adversely affecting research. U.S.-educated foreign nationals are going back to their home countries because of the difficulty of going through the process to stay in the United States.
20. It is also difficult for an individual to obtain a visa to enter the United States.
21. National assistance for nanotechnology development in foreign countries is more effective than in the United States. It will be a problem for competitiveness.
22. Some academics and researchers fight efforts for commercialization.

"Barriers to Nanotechnology Commercialization,"
a 2007 report for the U.S. Department of Commerce,
developed by the University of Illinois at Springfield

2

Types of Nanobusinesses

The mountain of information on nanotechnology is growing daily. Knowing where to go to get the information you need is key. This chapter focuses on key products and where you can go for additional research on products of interest. This chapter visits various websites and data sites to enable readers to know where to look to rapidly glean the information without months of research. Part 2 of this text offers several pages of Internet resources to increase your understanding of commercialization nanotechnology.

Small Times is a U.S. technology magazine that features daily articles covering nanotechnology, microelectromechanical systems (MEMS), and microsystems through a business point of view covering news, stocks, and the latest nanoproducts.

Small Times recently featured the top 10 nanotechnology universities.

1. University at Albany–the State University of New York (SUNY)
2. Cornell University
3. University of Michigan
4. Rice University
5. University of Pennsylvania
6. University of Virginia
7. University of North Carolina
8. Ohio State University
9. Northwestern University
10. University of Minnesota

The top 10 was compiled from a survey including questions about facilities, funding, courses, degrees, research programs, publishing, patenting, company formation, industrial partnerships, and more.

With more than $6.5 billion in public and private investments, College of Nanotechnology Science and Engineering (CNSE's) Albany NanoTech Complex has attracted over 250 global corporate partners—and is the most advanced research complex at any university in the world.

The University of Albany College of Nanotechnology Science and Engineering (CNSE) was created in 2004. It was the first college in the world dedicated to global education, research, development, and technology, and specifically to the deployment in the emerging disciplines of nanoscience, nanoengineering, nanobioscience, and nanoeconomics. By 2007 it was ranked the world's number one college for nanotechnology and microtechnology in the annual college ranking by *Small Times*.

Its goal is to prepare the next generation of scientists and researchers in nanotechnology through its "800,000 square foot complex that houses the most advanced 200mm/300mm wafer facilities in the academic world, including over 80,000 square feet of Class 1-capable cleanrooms equipped with 300mm wafer processing tools. The complex incorporates state-of-the-art, R&D and prototype manufacturing infrastructure for nano/microelectronics, nanophotonics and optoelectronics, nano/micro systems (MEMS) and nanopower science and technology."

The complex will provide the facilities and knowledge base to help take the U.S. initiatives for nanotechnology into a globally competitive commercial entity.

As a disruptive technology, nanotechnology holds the power to touch every industry at the molecular level. Often evolutionary, a disruptive technology is a new technological innovation, product, or service that is significantly cheaper than current products, higher performing, has greater functionality, and is more convenient to use, eventually overturning the existent technology or product that is leading the market.

For many individuals, as well as organizations, it is going to require a paradigm shift, a way of looking at and understanding the current technology and products in use. It will require a new understanding that the technology will affect all industries, requiring organizations to face change or fall behind and disappear if they fail to adapt. The understanding being that nanotechnology is the art and science of manipulating matter at the nanoscale (down to 1/100,000 the width of a human hair) to create new and unique materials and products.

Nanotechnology has been called the builder's final frontier. While most use *nano* to encompass the understanding of nanotechnology, *nano* in scientific jargon is a prefix meaning something extremely small, precisely 1/1,000,000,000 (one billionth) of a meter. *Nano* originally comes from the Greek word *nanos*, and from the Latin *nano*, and both words mean *dwarf*. This distinction is important to the scientific community because prefixes that are less than 1, such as milli (1/1,000) and micro (1/100,000), are Latin. The Greek prefixes are greater than 1, such as kilo (1,000), mega (1,000,000), and giga (1,000,000,000).

The smallest measurement known is a picometer (pm), which measures one trillionth of a meter (1/1,000,000,000,000). The picometer is used almost exclusively with particle physics and quantum physics. Atoms range from 62 to 520 pm in diameter. There is nothing to build at the pico scale, and it is only used in nuclear technology such as reactors and bombs, while nanotechnology impacts many technologies, as it works with the elements. The elements of the universe are fixed in number as seen in the periodic table (Figure 2.1).

Measurement Scale

1 meter	3.28 feet		Macroscale	Seen via human eye
1/100 meter	1 centimeter (cm)	$(1 \times 10^2 \text{ m} = 1 \text{ cm})$	Macroscale	
1/1,000 meter	1 millimeter (mm)	$(1 \times 10^3 \text{ m} = 1 \text{ mm})$	Macroscale	Seen via optical microscope
1/1,000,000 meter	1 micrometer (micron)	$(1 \times 10^6 \text{ m} = 1 \text{ μm})$	Microscale	Larger objects via optical microscope Smaller objects via electron microscope
1/1,000,000,000 meter	1 nanometer (nm)	$(1 \times 10^9 \text{ m} = 1 \text{ nm})$	Nanoscale	Larger objects via electron microscope Smaller objects via atomic force microscope
1/10,000,000,000	1 Angstrom	$(1 \times 10^9 \text{ m} = 1 \text{ Å})$	Angs scale	
1/1,000,000,000,000 meter	1 picometer (pm)		Picoscale	The pica scale is the size range of individual atoms

Note: Electron and atomic force microscopes use computer processing to create an image that the human eye can see.

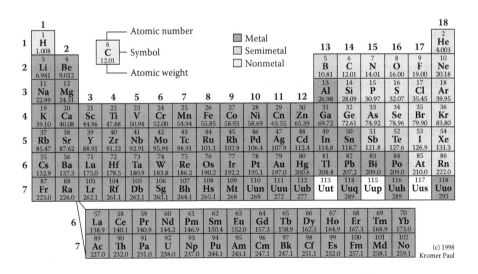

FIGURE 2.1

The periodic table of elements is the building block of everything created. Nanotechnology takes the building blocks to a new level at the nanolevel.

While it is exciting to explore the further reaches, for this book we will dwell on the commercialism of nanotechnology because the nanoscale impacts many technologies and can be used on a commercial basis. It has been called, by 1996 Nobel laureate from Rice University, Richard Smalley, "the builder's final frontier," and putting together atoms and molecules at the nanoscale resembles building with Lego blocks on a miniscule level.

In looking at the dimensions of a nanometer, why is size so important? At the nanoscale we are finally able to view such things as DNA, viruses, proteins, drug molecules, and quantum dots, whereas before, at the microscale, we could view human cells and bacteria, and on the macroscale we could look at tissues, human hairs, etc.

People first hearing about nanotechnology often think that it is a relatively new technology, when in reality it has been used for over 2,000 years, but we did not know it because we did not have the powerful electron and tunneling microscopes to see what was being created at that scale. Today, the scanning probe tools and microscopes convert the data, using computers, into images that we can see on the nanoscale and understand. Individuals can get a feel for what it is like to operate a transmission electron microscope (TEM) by going to http://nobelprize.org/educational_games/physics/microscopes/tem/tem.html.

The use of these magnificent microscopes allows us to look at products made in the past and discover that nanotechnology, while not known as such, was used in many products early on, such as stained glass and the legendary Arabian swords that had an almost magical strength (Figure 2.2).

The power that the scanning probe and tunneling microscopes has given us is the capacity to understand the nanoscale and to controllably and repeatedly create things (manufacture by any other name) at the nanoscale. With this power to manufacture products, we now have the capability to take them to a commercial scale.

Just as the powerful equipment can help us to understand some of the past, it also projects our future. *BusinessWeek* foreign correspondent Stephen Baker often writes about nanotechnology. In an interviewed with nanoscience pioneer Professor Chad Mirkin, who was named to President Obama's Science and Technology Advisory Council in 2009, Mirkin told Baker, "Nanoscience is about redoing everything. Everything when miniaturized will be new. Each day, new and exciting discoveries come out of the labs on how to work on the nanoscale, creating a multitude of new applications and accomplishments that benefit mankind. The entire industrialized world has begun to see the effects in pharmaceuticals, manufacturing, electronics, computers, defense, environmental science, communications and energy production."

The fact that the science is ready and available to enhance so many industries is exciting for growth it can make in these companies, but equally exciting with respect to the number of jobs it stands to create.

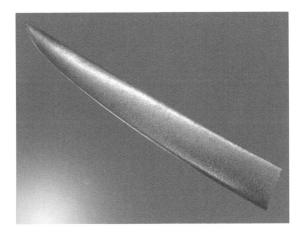

FIGURE 2.2
Cross section of Damascus sword. One of the oldest known (1100 A.D.) uses of carbon nanotubes and nanowires is in the legendary steel swords from Arab craftsmen in ancient Damascus. They had carbon nanotubes and nanowires in them. (From Reibold, M., et al., *Nature*, 444, 2006.)

Nanotechnology forecasts show that by 2015:

- Fifteen percent of the global manufactured goods will incorporate nanotechnology, giving the economy a boost of approximately $3 trillion in that market.
- Fifty percent of new technology products will incorporate nanotechnology.
- Two million nanotechnology workers will be needed worldwide.
- Five million support jobs will be generated to create the infrastructure of the industry.

This insight into where we are going in the near future will help economists, industry experts, businessmen, investors, politicians, students, and others interested in the emerging industries brought about by nanotechnology, in both the scientific realm and support businesses.

It is imperative that both individuals and companies understand that nanotechnology has the potential to change society as we now know it. Nearly $9 billion a year in research and development investments around the world is being poured into this new technology that will lead to new and more efficient energy production, storage, and transmission, more effective pollution reduction and prevention, medical treatments and tools, and lighter, stronger materials (Figure 2.3).

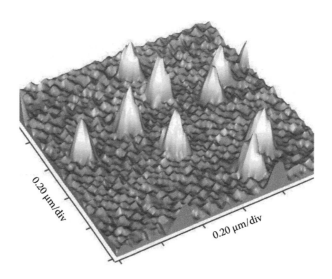

FIGURE 2.3
Three-dimensional image of gold nanodots. Colloidal gold is a suspension (or colloid) of sub-micrometer-sized particles of gold in a fluid. Due to the unique optical, electronic, and molecular recognition properties of gold nanoparticles, they are the subject of substantial research, with applications in a wide variety of areas, including electron microscopy, electronics, nanotechnology, and materials science. Often we would like to know the size of these nanodots. Ideally, we like for them to have the same size. Again, this can tell us how good the chemical recipe is (uniformity nanodot size, estimated quantity of nanodots in the solution). (Courtesy of The Pennsylvania State University.)

Keeping abreast of emerging technologies requires that businesses look beyond where the industry is moving because it changes so rapidly. Keeping tabs on the projection of futurists who study trends and emerging technologies will help business owners develop their businesses in keeping with the world needs. There are multiple emerging technology conferences around the world that allow business owners to network with thousands of business leaders, policy leaders interested in the growth of the industry, entrepreneurs, venture investors, and others interested in learning more about critical innovations rapidly emerging on the business radar in industries such as energy, biotechnology, IT, etc., to help understand how the up-and-coming trends in these areas will impact business. The conferences are held around the world on multiple industries. For the best results in finding one that suits your needs, do a web search of emerging technology conferences, adding the year interested and specifics to the industry.

For those entrepreneurs, investors, and business people looking toward trends when setting up their strategies for the next leg of their nanotechnology developments, the World Future Society (www.wft.org) predicts nanotechnology breakthroughs spanning the next 15 years.

1. Two to 5 years from now:
   a. Tires that only need air once a year
   b. Complete medical diagnostics on a single computer chip
   c. Go-anywhere concentrators that can produce drinkable water from air
2. Five to 10 years from now:
   a. Powerful computers you can wear or fold into your wallet
   b. Drugs that turn AIDS and cancer into manageable conditions
   c. Smart buildings that self-stabilize during earthquakes or bombings
3. Ten to 15 years from now:
   a. Artificial intelligence so sophisticated you can't tell if you are talking on the phone with a human or a machine
   b. Paint-on computer and entertainment video displays
   c. Elimination of invasive surgery, since bodies can be monitored and repaired almost totally from within

For those with a visionary perspective, the World Future Society has 10 several forecasts for the long range of 2030. Listed below are those specific to nanotechnology.

Everything you say and do will be recorded by 2030. By the late 2010s, ubiquitous unseen nanodevices will provide seamless communication and surveillance among all people everywhere. Humans will have nanoimplants, facilitating interaction in an omnipresent network. Everyone will have a unique Internet protocol (IP) address. Since nano storage capacity is almost limitless, all conversation and activity will be recorded and recoverable. (Gene Stephens, "Cybercrime in the Year 2025," *The Futurist*, July–August 2008)

Bioviolence will become a greater threat as the technology becomes more accessible. Emerging scientific disciplines (notably genomics, nanotechnology, and other microsciences) could pave the way for a bioattack. Bacteria and viruses could be altered to increase their lethality or to evade antibiotic treatment. (Barry Kellman, "Bioviolence: A Growing Threat," *The Futurist*, May–June 2008)

Careers, and the college majors for preparing for them, are becoming more specialized. An increase in unusual college majors may foretell the growth of unique new career specialties. Instead of simply majoring in business, more students are beginning to explore niche majors such as sustainable business, strategic intelligence, and entrepreneurship. Other unusual majors that are capturing students' imaginations: neuroscience and nanotechnology, computer and digital forensics, and comic book art. Scoff not: the market for comic books and graphic novels in the United States has grown 12% since 2006. (*The Futurist*, "World Trends and Forecasts," September–October 2008)

The World Future Society is comprised of 25,000 individuals around the world (80+ countries): business leaders, policy makers, educators, entrepreneurs, students, and retirees from virtually every field and industry.

While this may all seem like sci-fi from the latest Dean Koontz novel, there are a number of products already on the market and being used by many who don't even know they are using the new technology. A good first step in seeing what is out there for the layperson is to visit PEN—Project on Emerging Technologies (http://www.nanotechproject.org).

The Project on Emerging Nanotechnologies (PEN) is located in the Woodrow Wilson International Center for Scholars, located in Washington, D.C. The center is a nonpartisan institution that is supported by public as well as private funds is dedicated to the study of national and world affairs. Named for President Woodrow Wilson it is charged with strengthening relations between the world of learning and public affairs. By encouraging contacts and dialogue among policy makers, business leaders, and scholars, the center keeps the American public informed through various multimedia sources.

PEN was launched in 2005 by the Wilson Center and the Pew Charitable Trusts. It is dedicated to helping businesses, governments, and the public anticipate and manage the possible health and environmental implications of nanotechnology. For more infomation about the Project on Emerging Nanotechnologies visit http://www.wilsoncenter.org/nanotech or http://www.nanotechproject.org.

PEN has an inventory database of current nanoproducts in use:

http://www.nanotechproject.org/inventories

http://www.nanotechproject.org/inventories/consumer/analysis/

http://www.nanotechproject.org/inventories/medicine/apps/imaging/trilite_technology/

This is an "essential resource for consumers, citizens, policymakers," and others interested in monitoring how nanotechnology is entering the marketplace and the products it is bringing to the world table, companies producing those products, and the countries involved in the development. The site is dynamic and encourages users to submit updated and new information to nano@wilsoncenter.org.

Listed below is a breakdown of the eight categories provided by the PEN database:

1. **Consumer products.** This section contains over 1,014 products, produced by 484 companies that are currently located in 24 countries (Figure 2.4).

2. **Environment, health, and safety research.** This section showcases 561 environmental, health, and safety projects that are located in 17 countries, featuring an international and expanding inventory. See Figure 2.5 for an example.

3. **U.S. nano metro map.** The map section of the database shows the location (by zip code) of companies, universities, government laboratories, and organizations in the United States that are working on nanotechnology. The U.S. nano metro map shows areas that have more than 15 universities, government labs, companies, or organizations. The top nano metros are Boston, MA; San Francisco, CA; San Jose, CA; Raleigh, NC; Middlesex-Essex, MA; Oakland, CA; San Diego, CA; Seattle, WA; and Austin, TX.

FIGURE 2.4
Wilson™ double-core tennis balls (http://www.wilson.com, http://www.racketsdirect.com/).

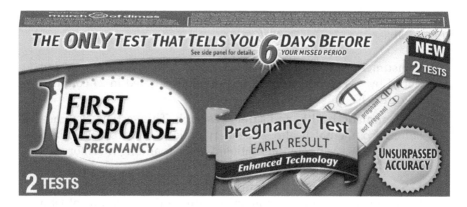

FIGURE 2.5
First Response® home pregnancy test. "First Response home-pregnancy test manufactured by Carter-Wallace, a New York-based biotechnology company. The test uses gold particles (less than 50 nanometers in diameter) to help consumers read test results more easily" (*Christian Science Monitor*, 2009).

FIGURE 2.6
Synthetic biology map. It is estimated that by 2015, one-fifth of the chemical industry ($1.8 trillion) could be dependent on synthetic biology. The United States is currently the world leader in this emerging science.

4. **Synthetic biology map.** An interactive map of synthetic biology activity across the United States and Europe. Synthetic biology is "the convergence of molecular biology, information technology, and nanotechnology, leading to the systematic design of biological systems" (KNAW—Rathenau Institute in Netherlands) (Figure 2.6).

FIGURE 2.7
Worldwide nanotechnology food market, $20.4 billion by 2010.

5. **Agriculture and food.** The Agrifood Nanotechnology Research and Development Database posits that the worldwide nanotechnology food market will be $20.4 billion by 2010, with 5 out of 10 of world's largest beverage companies already investing in nanotechnology R&D. As our world food resources change, it is critical to utilize the developments of nanotechnology, thus pushing commercialization of nanotechnology in this market (Figure 2.7).

6. **Medicine.** Nanotechnology medical developments over the coming years will have a wide variety of uses and could potentially save a great number of lives, but this too demands commercialization timelines that can broach the valley of death, getting the technology from the labs to the market. One of most immediate related medical research developments focusing on cancer currently has data that include 130 nano-based drugs and delivery systems and over 125 devices and/or diagnostic tests.

EXAMPLE: QUANTUM DOTS

Quantum dots (QDs) are being studied for use in tracking cancer cell movement in the lymph system (Figure 2.8). The dots are injected into a tumor and their exodus to a lymph node is observed by watching the color they reemit. This enables doctors to pinpoint lymph nodes that contain the cancer cells.

FIGURE 2.8
Quantum dots. Energized by light, quantum dots reemit light or fluoresce. The color they reemit depends on the size and material composition of nanoscale quantum dots, enabling them to be used in tumor tracking. (From Evidenttech.com.)

7. **Silver nanotechnology.** The Silver Nanotechnology in Commercial Products Database shows nanosilver applications available on the market that have potential to contact or affect the public. Nanosilver technology eliminates bacteria from products it is applied to, promoting a germ-free environment. This database on silver nanotechnology includes consumer products, medical applications, and precursor products that will be incorporated into final products that claim to use silver nanotechnology or silver colloids. The database has over 244 products that contain silver materials, including items such as food containers, nanosilver cutting boards, sunscreen, fabric softener, cotton sheets, bath and sport towels, nanosilver dust power, mineral supplements, nanosilver mesh filtration, hair appliances, etc. (Figure 2.9).

8. **Remediation maps.** The remediation maps show the locations of contaminated sites utilizing some form of nanoremediation. Sites include oil fields, manufacturing sites, military installations, private properties, and residences. The 2004 Environmental Protection Agency (EPA) estimates that it will take up to 35 years and cost up to $250 billion to clean up the nation's hazardous waste sites. Nanoremediation has potential to cut costs and reduce cleanup time, but ecotoxicity of nano is complex and there is not enough data on how it will affect human health. The sites are in seven countries, including the United States, with only a fraction of the projects reported, and new projects show up regularly.

Nanostrategies is a full service consulting division of Nanotechnology Now. It recently published a report of the industries that nanotechnology will likely have a *disruptive* effect on in the near term (amounts are billions of U.S. dollars):

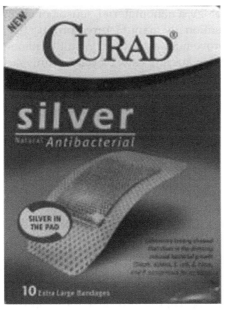

FIGURE 2.9

Curad® Silver Bandages. Nanosilver technology eliminates bacteria from products to which it is applied, promoting a germ-free environment. (From http://www.curadusa.com.)

- $1,700 healthcare
- $600 long-term care
- $550 electronics
- $550 telecom
- $480 packaging
- $450 U.S. chemical
- $460 plastics
- $182 apparel
- $180 pharmaceutical
- $165 tobacco
- $100 semiconductor
- $92 hospitality/restaurant
- $90 U.S. insurance
- $80 corrosion removal
- $83 printing

- $70 mouthwash
- $57 U.S. steel
- $43 newspaper
- $42 diet supplement
- $40 diet
- $32 publishing
- $30 catalysts
- $27 glass
- $24 advertising
- $18 cosmetics
- $13 chocolate
- $10 battery
- $5 blue jeans
- $4 khakis
- $2.8 fluorescent tagging

According to the Project on Emerging Nanotechnologies, the number of nanoproducts available to the public has topped the 1,000 mark. The majority of these are conventional products that include some component created by nanoscale technology, a nanomaterial, sensor, or chemical integrated into a much larger conventionally manufactured product.

Performance improvement as a direct result of nanotechnology is being seen in today's military aircraft, cars, bicycles, tennis rackets, golf clubs, etc. Whatever your passion, nanotechnology is finding a way to enhance the product. No idea seems to be out of the realm of potential for this technology in the future. How about a book made of electronic paper that contains miniature pixels of self-assembled monolayers of nanoscale materials?

The clothing industry is rapidly getting on board with clothing that cleans itself when the wearer exposes it to sunlight (Wang, 2008). The Chinese have been pouring billions into nanotechnology enhancing their products to make them more globally appealing, thus stimulating their already progressive economic strategy. A fairly recent discovery was made by the Chinese Academy of Sciences (CAS), who created a surface disposal technique for fabrics such as silk, wool, and cotton to make them waterproof and oil-proof. A line of "nano neckties" was created to shed the eating stains that ruin so many expensive ties. The scientists created the surface disposal technique after watching the lotus leaf as water slid off the leaf, not absorbing the moisture. Just as on the surface of the lotus leaf, the scientists developed a "fuzz-like framework on top of the tie, just like that on the lotus leaves, with a size smaller than 100 nano-meters," said Dong Yongrong, general manager of the necktie producer who partnered with the CAS on the project. "The fuzz absorbs air molecules and forms a thin covering that protects the cloth from oil and water." This nanotreatment left the materials retaining the soft qualities of the silk and cotton as opposed to traditional waterproofing techniques that made the silk feel rough.

The scientists at CAS are looking toward the future with the textile industry, foreseeing clothes of the future that can "sense the surrounding changes of light, sound, temperature, humidity, radiation, and even fluctuations in body temperature."

On a more personal side for men, nanotechnology has come to the rescue for millions of men who also experience erectile dysfunction (ED), with heart trouble and other health problems affecting erectile function. The article published in an online edition of the *Journal of Sexual Medicine* noted that tens of millions of men around the world have benefited from oral ED medications, but noted that these drugs include side effects like headache, facial flushing, nasal congestion, upset stomach, abnormal vision, as well as isolated reports of hearing and vision loss. Men who have suffered heart attacks or strokes have been cautioned to use the medications with caution or not at all. In addition, "an estimated 30-to-50 percent of men with ED do not respond to oral use of PDE5 inhibitors," says senior author Kelvin P. Davies, PhD, associate professor of urology at the Albert Einstein College of Medicine.

Einstein scientists have developed a new treatment for ED—"a drug delivery system consisting of nanoparticles, each smaller than a grain of pollen, that carry tiny payloads of various drugs or other medically useful substances and release them in a controlled and sustained manner." The treatment is a topical therapy rather than oral medication and can be used by men, particularly those with diabetes, who have reduced levels of nitric oxide.

Albert Einstein College of Medicine of Yeshiva University is one of the nation's premier centers for research, medical education, and clinical investigation. It is home to 2,775 faculty members, 625 MD students, 225 PhD students, 125 students in the combined MD/PhD program, and 380 postdoctoral research fellows. In 2008, the Albert Einstein College of Medicine received more than $130 million in support from the NIH. This includes the funding of major research centers at Einstein in diabetes, cancer, liver disease, and AIDS. The College of Medicine is also focusing its efforts in developmental brain research, neuroscience, and cardiac disease.

The Albert Einstein College of Medicine has an extensive affiliation network involving eight hospitals and medical centers in the Bronx, Manhattan, and Long Island—which includes Montefiore Medical Center, the University Hospital, and Academic Medical Center for Einstein. The College of Medicine runs one of the largest postgraduate medical training programs in the United States, offering approximately 150 residency programs to more than 2,500 physicians in training.

Around the world, new products are brought to the market daily. The list is extensive and there is no way to feature them all. Following are just a few nanotechnology companies from around the world to show a cross section of countries involved in nanotechnology and some of their efforts:

- **Shenzhen, China—Chengyin Technology:** Focused on R&D of nanostructured titanium dioxide in sunscreen, antimicrobial, antistatic, and photocatalysis.

- **Taiwan—Asia Pacific Fuel Cell Technology:** PEM fuel cells with technologies in stack design and manufacturing, system integration, and metal hydride hydrogen storage.

- **Selangor, Malaysia—Nanopac:** Photocatalytic/functionalized materials, processing technology for powders, coatings, and composites, control of surface modification, analysis/evaluation/optimization of materials and processing, and design of components and systems (air purification, water purification).

- **New South Wales, Australia—CAP-XX:** Develops supercapacitors—high-power, high-energy storage devices that enable manufacturers to make smaller, thinner, and longer-running products such as mobile phones, PDAs, medical devices, AMRs, compact flash cards, and more.

- **Edmonton, Canada—Quantiam Technologies:** Develops and commercializes advanced materials based on nanotechnology; manufactures powders, catalysts, and coatings for the petrochemical, energy, and aerospace industries; provides consulting, technical, and research services for characterization of nanomaterials and surfaces; extensive collaboration and seed investment in innovation.

- **Schwarzenau, Austria—Metcomb Nanostructures:** With its patented Integrated Nanostructure Control process, Metcomb leverages nanotechnology to create the first aluminum foam that is uniformly consistent, mirroring the way mother nature creates organic load-bearing materials, such as wood, bone, and coral.

- **Lahti, Finland—Amroy:** The latest product family is Hybtonite®-nanoepoxies. Hybtonites can be used in marine, automotive, wind energy, and in many industrial applications.

- **Grenoble, France—Nanopolis:** Whether you are a researcher, industrialist, or educator involved in the emerging field of nanotechnology, Nanopolis multimedia-distributed knowledge network and encyclopedia series provides you a straightforward way to understand the nanotech world and to be understood within.

- **Berlin, Germany—Capsulution NanoScience AG:** A Berlin-based company, which has started with a vision: to create innovative, highly specialized products by applying its unique LBL-Technology® to existing and emerging needs in life science developments. The LBL-Technology (derived from the term *layer by layer*) is a high-tech tool for making unique multifunctional nano- and micron-sized capsules that are invisible to the human eye.

- **Heusweiler, Germany—Nanopool:** The German Ministry of Research and Education promotes the project Ultra Thin Layers as being one of seven clearly defined disciplines of nanotechnology. The manufacturing of those ultra-thin nano layers is the core competency of Nanopool GmbH.

- **Gibraltar—Cool Chips:** The future of all cooling, refrigeration, and thermal management. Cool Chips uses a revolutionary new thermo-tunnel technology to deliver up to a projected 70–80% of the maximum (Carnot) theoretical efficiency for heat pumps.

- **Athens, Greece—Acrongenomics:** A publicly traded nanobiotechnology company that specializes in the field of research and development of solutions in the fields of genomics, proteomics, and diagnostics.

- **Dublin, Ireland—Ntera:** A broad-based nanotechnology company with current applications in flat-panel displays, medical diagnostic sensors, and targeted drug delivery.

- **Amsterdam, Netherlands—Agendia:** Developed high-quality methods, using microarray genetic profiling, for analyzing tumor samples and mapping the tumor's specific properties.

- **Madrid, Spain—CMP Cientifica:** Europe's first integrated solutions provider for the nanotechnology community, specializing in providing nanotechnology information to the scientific and financial communities; linking science and industry through networks of excellence and conferences; providing expert solutions for high-technology and advanced manufacturing companies; and venture capital funding for nanotechnology start-ups.

- **Lund, Sweden—Nanofreeze:** Founded in September 2005, as a spin-off from research performed at Lund University in the field of thermoelectrics.

- **Bristol, UK—NanoMagnetics Ltd.:** Developing advanced magnetic materials for the information storage industry using its patented protein-based technology. These materials have the potential to replace the magnetic thin-film technology used today by the hard disk industry, which is seeking to overcome the approaching physical superparamagnetic storage density limit.

- **London, England—JR Nanotech:** A healthcare company that promotes the application of nanosilver. It uses nanosilver particles as an antibacterial substance to provide effective and complication-free treatment for age-old healthcare.

- **Manchester, England—Nanoco Technologies:** Develops and manufactures fluorescent nanocrystals from semiconductor and metallic materials known as quantum dots.

- **Oxford, England—Oxford Nanotechnology:** A research and development company based in Oxford. Its research specialty is the use of ion beams to create nanostructures, but it also has interests in nanoscale electronic architecture, quantum computing, and other nanotechnology research. It is currently looking to expand its operations and employ a number of staff research scientists.

- **Oxfordshire, England—P2i:** P2i's plasma surface enhancement process is a 21st-century technology that gives everyday products extraordinary performance levels of oil and water repellency.

- **Hertzelia, Israel—NutraLease:** The technology is a patent pending called Nano-sized Self-assembled Liquid Structures (NSSL) and is related to the nano-sized vehicles that are used to target compounds (such as nutraceuticals and drugs).

- **Moscow, Russia—Invest Technologies:** The enterprise takes part in the activity of Russian Committee ultra-fine (nano) materials for production, research of properties, and application of UFP. It also cooperates with leading scientific institutes, conducting studies of nanocrystal materials.

- **Sao Paulo, Brazil—Center for Automation in Nanobiotech (CAN):** Focuses on investigation of new paradigms for innovation in systems and automation design. CAN's main thrust and aim is the development of practical and useful nanobiotechnology systems and devices that may benefit people around the globe with biomedical engineering advances.

- **Monterrey, Mexico—NanoMerk:** Distributes and commercializes high-technology products that offer a different and innovative solution to everything available in the market.

- **Huntsville, Alabama—MEMS Optical:** Supplier and manufacturer of both refractive (micro lens arrays) and diffractive (beam shapers, beam splitters, etc.) micro optics, and of MEMS devices such as scanning tilt micro mirrors and deformable mirrors.

- **Costa Mesa, California—QuantumSphere:** A leading manufacturer of metallic nanopowders for aerospace, defense, energy, and other markets.

- **Los Angeles, California—Abraxis Oncology:** The proprietary drug division of American Pharmaceutical Partners. Dedicated to improving treatments for patients with cancer. Maker of Abraxane—paclitaxel protein-bound nanoparticles for injectable suspension.

- **San Francisco, California—Optiva:** Benefiting from over two decades of successful large-scale research in supramolecular engineering, Optiva is the first to develop a commercial mass production process of optical self-assembling materials (nanomaterials). Optiva's first family of thin crystal film (TCF™) products will target the flat-panel display industry, offering significant cost and performance advantages over traditional polarizer film alternatives.

- **Santa Monica, California—RAND:** RAND (a contraction of the term *research and development*) is the first organization to be called a think tank. Its job is to help improve policy and decision making through research and analysis.

- **Santa Clara, California—Intel:** Intel supplies the computing and communications industries with chips, boards, systems, and software building blocks that are the "ingredients" of computers, servers, and networking and communications products. Intel is behind everything from the fastest processor in the world to the cables that power high-speed Internet.

- **Broomfield, Colorado—ALD NanoSolutions:** Commercialization strategy is founded on the assertion that new materials will be designed rather than discovered. The compelling opportunity is to identify and synthesize a new set of composite materials that are comprised of common substrates coated with specific material.

- **Farmington, Connecticut—US NanoCorp (USN):** USN was incorporated April 1, 1996, as a vehicle to identify, develop, and commercialize value-added products in the field of energy storage and energy conversion devices that exploit the extraordinary properties of nanostructured materials.

- **Coral Gables, Florida—Radiation Shield Technologies (RST):** Proud to offer Demron™, the new standard in personal radiation protection. This revolutionary technology is currently produced as full body suits, gloves, and boots.

- **Chicago, Illinois—NanoInk:** In business to commercialize the process known as dip pen nanolithography™, or DPN™. We have developed a platform process for nanotechnology that enables our customers and partners to create revolutionary new products and services using a molecular "bottom-up" approach to nanofabrication.

- **New York, New York—Owlstone Nanotech:** Using leading-edge nanofabrication techniques, Owlstone has created a complete chemical detection system on a chip. A hundred times smaller—and a thousand times cheaper—than other currently available devices, the Owlstone detector overcomes many of the limitations of traditional detection technologies.

- **Chapel Hill, North Carolina—Xintek:** Formerly Applied Nanotechnologies, Inc. (ANI). Established in October 2000 to develop and commercialize applications of carbon nanotubes in various industries such as telecommunications, electronics, and medical imaging systems. The company also fabricates carbon nanotubes for use in the above-mentioned industries and the research community.

- **Research Triangle Park, North Carolina—Research Triangle Institute (RTI):** In 2001, RTI initiated a new focus on nanotechnology to consolidate and coordinate years of successful work in thermoelectrics, materials science and engineering, and filtration and aerosol technology.

- **Research Triangle Park, North Carolina—Alnis BioSciences:** A drug development company with a potent, enabling therapeutic platform to treat cancer as well as infectious and inflammatory diseases. Alnis has engineered nanoscopic hydrogels, or NanoGels, comprised of polymers, bioactives, and targeting molecules.

- **State College, Pennsylvania—NanoHorizons:** A nanoscale materials and devices company providing solutions for drug discovery, flexible microelectronics, and medical diagnostics and monitoring.

- **Addison, Texas—Authentix:** Formed by the merger of Isotag, Biocode, and Calyx in 2003, Authentix combines 20 years experience in providing comprehensive services and technology for the prevention of product counterfeiting, brand adulteration, and product diversion.

- **Dallas, Texas—Raytheon's Nanoelectronics Branch:** Develops future generation analog and digital technologies for commercial and defense applications in radar, communications, and sensor processing.

- **Blacksburg, Virginia—Luna nanoWorks:** A division of Luna Innovations Inc. Luna nanoWorks' new composition of matter is called Trimetasphere™ carbon nanomaterials (M3N@C80)—a fullerene sphere enclosing three metal atoms in a nitride molecule. The entrapped metals provide unique physical, chemical, thermal, magnetic, biological, optical, and electronic properties that differentiate them from other carbon nanomaterials.

So how can these products come to life in a tightened economy with the valley of death facing new innovations? In the last chapter we identified how nanocenters can help with funding, research, and support to the new business. We will take the center concept a step further to look at some of the business applications that the centers have helped to bring to commercialism, as well as look at other specific industries that feature different applications for nanotechnology.

The National Nanotechnology Infrastructure Network (NNIN) is a partnership of 13 user facilities, supported by the National Science Foundation (NSF), providing opportunities for nanoscience and nanotechnology research.

The following products are a few listed in the NNIN resource section, and while the NNIN does not endorse any of the manufacturers or suppliers of the products, it is providing courses for obtaining materials used in their Exploring Nanotechnology through Consumer Products unit.

Behr® Premium Plus Kitchen and Bath Paint
http://www.behr.com
Home Depot and other hardware and painting supply stores

CD vs. DVD
Electronics stores, office supply stores, supercenters, etc.

Clarity® Fog Eliminator
http://www.nanofilmtechnology.com
http://www.amazon.com/
Sporting goods stores

DERMAdoctor® Poutlandish
http://www.dermadoctor.com/
http://www.softsurroundings.com/P/POUTlandish/
http://www.amazon.com/

Eagle One® Nano Wax
http://www.eagleone.com/
http://www.autobarn.com/
Most automotive supply stores and supercenters

Flex Power™ Joint and Muscle
http://www.flexpower.com/flex_work.html
http://www.amazon.com/
http://www.bodybuilding.com/

Head® "Nano Titanium" Tennis Racket
http://www.head.com/tennis/racquets.php?region=us
http://www.pwp.com/tennisrackets
Most sporting goods stores everywhere

Kodak® Ultima Picture Paper
http://www.kodak.com
Office and photo supply stores everywhere

L'Oreal™ Retivalift
Drugstores and supercenters everywhere

NanoBreeze® car air filter from NanoTwin Technologies
http://www.NanoBreeze.com/
http://www.automotiveinteriors.com/

Nano Mask®
http://www.nano2mask.com
http://www.nanobuddy.com
http://birdfluprotection.com/nanomasks/

NanoTex™ fabric
http://www.nano-tex.com
Nano-enhanced clothing is available at Dockers, Eddie Bauer, Old Navy, LL Bean, Land's End,
  Gap, and Lee Jeans; look for articles featuring stain-resistant fabric

NDMX™ golf balls
http://www.ndmxgold.com/
Sporting goods stores everywhere
Also, Wilson golf balls feature the NANO TECH™ CORE: Nanoparticles infused into the
  rubber chemistry produce an extremely lively but soft inner core
http://www.wilson.com

Pilkington ACTIV™ Self Cleaning Glass
http://www.pilkington.com/international+products/activ/usa/english/interested/default.htm
Building and window supply stores

Wilson™ Double Core Tennis Balls
http://www.wilson.com
http://www.racketsdirect.com/
Sporting goods stores everywhere

X-static™ Nano socks
http://www.x-staticfiber.com
http://www.drsocks.com/xstatic.shtml
http://www.duluthtrading.com

Curad® Silver Bandages
http://www.curadusa.com
Drugstores and supercenters everywhere

*Additional information on consumer products utilizing nanotechnology is available at http://www.
nanotechproject.org (a nanotechnology consumer products inventory). At this site you will find a list of
over 200 products with a description of the product and a link to the manufacturer's website.*

CASE STUDY: PITTSBURGH NANOMATERIALS COMMERCIALIZATION CENTER, PITTSBURGH, PENNSYLVANIA

The Pennsylvania NanoMaterials Commercialization Center was founded in 2006 to fund nanotechnology start-up companies that need initial monies to develop promising research into marketable products. It is considered by many struggling nanobusinesses a bridge over the harsh landscape of commercialization, otherwise known as the valley of death. The center is based in the technology corridor in Pittsburgh, Pennsylvania.

The Pennsylvania nanocenter focuses on new nano-based products for existing companies as well as helping with the formation and growth of new research concepts, taking them from the valley of death, which includes the stages of initial research, market needs, and the first prototype of the start-up company. As the company moves toward manufacturing, the Pennsylvania NanoMaterials Commercialization Center helps to develop and evaluate the prototype toward pilot productions and establish partnerships as the production is handed off to large companies, venture capitalists, and angels for scale-up to market.

"We are very early-stage capital investors where venture capitalists won't go," says Alan Brown, the PA nanocenter's executive director.

The center has invested in diverse companies that secured the support of the center by submitting a winning proposal during one of its funding rounds. Pennsylvania NanoMaterials Commercialization Center features a NanoMaterials Commercialization portfolio of the companies that it is helping to develop and commercialize (http://www.pananocenter.org/project-portfolio.aspx). The following is a look at some of the companies the center has helped and the business applications they created via nanotechnology.

1. **Metalon.** Metalon is a Carnegie Mellon University start-up company that was formed in January 2010 to use disruptive technology to further the field of printed electronics, thereby producing low-cost, printable, and disposable devices that cover a wide spectrum of technologies. Metalon supplies "molecular inks comprised of novel metal complexes that can be printed as either solutions or neat liquids. These materials can then metalize, thermally or photochemically, to form highly conductive traces and structures on a variety of substrates, including flexible organic supports."

2. **SolarPA, Inc.** SolarPA is working on Nanocoat—a nanocrystalline coating that bends incoming sunlight to minimize the reflection off solar panels, trapping light inside the

semiconductor material and redirecting the incoming light along the surface, increasing absorption rate, rather than having the light pass through a semiconductor material. The good news is that the technology is relatively inexpensive and is projected to lower the cost per watt of solar power.

3. **University of Pittsburgh** (precommercialization project). Researchers are developing a project for nanocharacterization by creating an *ex situ* environment reactor that is compatible with a transmission electron microscope (TEM) heating holder that can be inserted and exposed to specific gas reactants and pressures then placed in the TEM. This development is expected to impact various energy-related technologies.

4. **ICx Technologies, Inc.** (precommercialization project). ICx is working on a next-generation biofuel, developing a process for the production of chemicals that are found to be identical to diesel fuel, having the potential to be compatible with existing engines and compatible with multiple sources of energy, while independent of food-related energy sources so as not to compete with food production or increase food prices. ICx is a leader of advanced biotechnologies for both government and commercial applications.

5. **Pennsylvania State University** (precommercialization Project). Penn State University is working on a graphene-based nanocomposite with high energy density or high power in energy storage devices that can improve electrode kinetic and cycling stability for energy storage applications such as Li-on battery and supercapacitors.

6. **Industrial Learning Systems, Inc.** iLS is developing a silicon wafering technology that was patented by Carnegie Mellon University and licensed to iLS that will provide a continuous production of nanostructured solar cells. The process has the potential to "reduce wafer cost by a factor of 4 or more so that solar electricity will reach grid parity at about 40–60c/kW."

7. **Crystalplex Corporation.** By developing a proprietary quantum dot (DT) technology, Crystalplex technology now manufactures a more cost-effective and efficient solid-state light source at performance levels "beyond the reach of current rare-earth phosphor-based white LEDs," producing a QD-LED.

8. **Illuminex Corporation.** Illuminex is working on the commercialization of copper-wilicon nanostructured anode for lithium-ion batteries, leading to higher-energy-density LIBs used in portable electronics. This technology will also be significant in the development of electric vehicles.

9. **Kurt J. Lesker Company and Intergran Technologies USA.** These organizations are working in partnership for the commercial development of nCu (high-purity copper)-sputtering targets that are used in the fabrication of next-generation semiconductor devices using Nanovate™ technology developed by Integran Technologies USA. Combining the global opportunity and manufacturing experience of Kurt J. Lesker Company and Integran's advanced material technology, these organizations have created a unique global opportunity for both semiconductor and materials customers.

10. **NanoLambda, Inc.** NanoLambda develops ultra-compact, highly accurate LED color (wavelength) monitoring sensors with a 2 nm accuracy, combining the nano-optic filter array technology and monolithic nanoimprint process for a low-pilot production, monitoring color quality/consistency of LEDS over time and temperature, which appears to be the biggest challenge in the LED market.

11. **nanoGriptech, LLC.** nanoGriptech is a spin-off company of Carnegie Mellon University and has created a polymer fibrillar adhesive technology that mimics gecko foot hairs that allow them to grip repeatedly strongly on both smooth and rough surfaces in both wet and dirty outdoor conditions, perfect for materials and sporting goods industries.

12. **Arkema, Inc.** In partnership with Lehigh University, Arkema is developing a Nanostrength® block copolymer technology for toughening epoxies in wind energy and electronic materials applications that will increase the reliability of wind blades without sacrificing strength.

13. **Strategic Polymer Sciences, Inc.** Strategic Polymer Sciences is a spin-off company of Penn State University with an exclusive license of the electroactive polymer technique invented by Dr. Qiming Zhang. The company has developed an advanced nanostructured polymer hybrid capacitor film and prototype capacitors for implantable cardioverter defibrillators (ICDs), making them widely accessible for millions of Americans subject to sudden cardiac arrest. The capacitor has high density and reliability and can be produced in less than 3 mm and used in multiple applications, including medical devices, power electronics, hybrid electric vehicles, and military weapon systems.

14. **Bayer MaterialScience, LLC.** Bayer is developing flexible sensing films that can prevent pressure sores using Baytubes® carbon nanotubes. The project is in partnership with Quality of Life Technology Center, which was founded jointly by Carnegie Mellon University and University of Pittsburgh.

15. **Y-Carbon, Inc.** Y-Carbon is producing nanoporous carbon technology for high-energy-density and high-power-density supercapacitors for a variety of electrical energy storage and management applications, including electronics, automotive industry, and backup power.
16. **PlextronicsSM, Inc.** Plextronics has created a high-performing active layer technology for organic photovoltaic solar cells known as Plexcore™ PV that will develop a new generation of polymer-based inks that will increase solar conversion efficiency and extend the life of existing organic semiconductor devices.
17. **Integran Technologies USA.** Integran is developing a new nanomaterials coaxial wire technology increasing the durability of lightweight electrical wiring systems. "This project will help to establish the new wire technology as a fully proven, mass production-ready process and to create a new technology and market support center in Pittsburgh to serve the wiring industry across the U.S."
18. **HydroGen Corporation.** HydroGen is developing enhanced performance hydrogen fuel cell electrodes using carbon nanotubes to improve the performance and lifetime of the electrodes, reducing overall costs for fuel cell electrodes that will expand its fuel cell business worldwide.
19. **NanoRDC, LLC.** NanoRDC is working on a project using treated carbon nanotubes (t-CNT) for use as electrical conducting fillers with improved distribution into thermoplastics.

Take a look at one particular story of success: Y-Carbon. Y-Carbon is a start-up based in King of Prussia, Pennsylvania, that is creating commercial applications for porous carbon research from the Drexel University's Nanotechnology Institute.

Drexel grad student Ranjan Dash discovered a way to use metal carbides to control the pore sizes of carbon, tailoring the carbon's properties for specific uses, such as increasing the energy storage in supercapacitors and hydrogen storage. Dash founded the company in 2004, becoming its chief technology officer, but the company didn't become active until 2007. In 2008, the company applied for funding from the Pennsylvania nanocenter and received $250,000 to build a prototype supercapacitor with twice the usual storage for a similar production price.

With this seed money in hand to cover initial expenses, the company now is raising money to build an assembly line. Dash credits the Pennsylvania nanocenter with his growth via the mandatory milestones that the center focuses on and by taking at least 2 years off his development timeline. Y-Carbon is planning on participating in the Pennsylvania nanocenter's mentoring program, which links start-ups with large corporations.

The mentoring program relationship is a mutually beneficial program. Start-ups provide the inventions and innovative ideas, and interested big businesses provide the money to develop the products for market. A technology advisory committee of 20 members reviews the ideas, which must come from Pennsylvania companies. Over the past 3 years, the center has received nearly 100 proposals. The review is based on the funding that is needed, compared to the potential value. Once the review is complete the advisory committee submits the top proposals to the Pennsylvania nanocenter's board of directors. From there, companies chosen receive at least $150,000. So far, 17 projects from 15 companies have been funded.

The funding comes from a variety of sources, including the military, various foundational investments, and the Pennsylvania Ben Franklin Technology Partners.

Going forward, the center hopes to interact more with private investors. "We think we can help them by basically showcasing these companies so [investors] can have a much earlier look at what they're going to invest in," Brown says. "We would like to hand these small companies off to private investors."

Beyond that, the center is looking for more funding it can hand out because as technology advances, ever more ideas spring up that could be worth funding. "The impact that we have is obviously proportional to the size of the budget," Brown says. "We turn away more projects than we fund."

The Pennsylvania NanoMaterials Commercialization Center's executive summary states that it is a nonprofit technology-based economic development organization focused on bringing nanotechnology research into the marketplace. The center's three critical objectives are:

- Accelerating the transition of new nanomaterials technologies to the defense and commercial markets
- Meeting the needs of the center's diverse stakeholders with value-added service
- Operating a lean business model to ensure maximum benefit to its stakeholders (PA nanocenter)

Figure 2.10 defines the difference between trying to enter the nanotechnology market with or without the help of the Pennsylvania nanocenter and its resources.

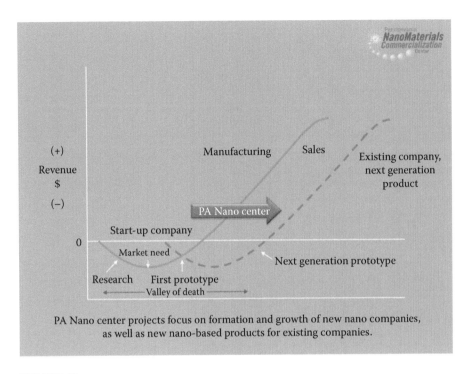

FIGURE 2.10
(From Pennsylvania NanoMaterials Commercialization Center, http://www.pananocenter. org/nano-center-about.aspx.)

3

Ease of Entry

There is a commonly held notion that there are four basic factors that are necessary for a successful business start-up venture: a profitable business idea, a thorough business plan, necessary and adequate capital, and a qualified entrepreneur.

Having the passion of the idea is not enough to take you to business success. Knowing if you are qualified to take on the trials and tribulations of business ownership, especially if you are not from a business background, takes a step into Entrepreneurial Education 101: Where do you stand with regard to being able to sustain a viable business enterprise?

The following is a listing of key characteristics from the Pennsylvania *Entrepreneur's Guide* for 2009 of successful business owners. See where you rank in terms of the traits you already possess and those that you may want to improve upon.

1. **Problem solving:** Can explore innovative ways to respond to opportunities.
2. **Goal oriented:** Can envision a desired outcome, as well as plan and implement the activities required to achieve it.
3. **Self-confidence:** Believes in own ideas and abilities, and conveys those beliefs to others.
4. **Risk taking:** Can abandon status quo, explore options, and pursue opportunities.
5. **Decision making:** Ability to make prudent choices even in a stressful environment.
6. **Persistence:** Can tenaciously pursue goals regardless of the energy and commitment required.
7. **Communication:** Can speak, listen, and write effectively.
8. **Interpersonal relationships:** Can understand the wants and needs of others, as well as inspire them.
9. **Leadership:** Can direct others effectively and empower their performance.

The guide followed this list by saying:

> As an entrepreneur you must possess not only the personal traits for success, you must also possess some degree of expertise in each of the managerial skills required for business survival and growth. Although you can hire skilled employees, engage qualified consultants, and develop a corporate advisory board, ultimately, it is you who must determine the strengths and weaknesses of your business, diagnose problems and seek out the expertise of others. You must learn to wear each of the managerial "hats," sometimes all at one time. Entrepreneurial tasks include the supervision and even performance of financial management, personnel management, marketing management and production management. If you have not developed the experience necessary to learn the basic skills of each of these areas, begin now to build them as a mandatory foundation for your business.

Great advice and one that would have saved countless dot-com companies who started with a genius idea and few, if any, business skills. Many successful owners rode the wave of money and success in the beginning but lost it all when things tightened and they found that they did not have the business skills to sustain growth and survival. This is an important lesson to many scientists looking to capitalize on their ideas in nanotechnology. As quoted earlier, you can hire what you don't know, but you have to be articulate and savvy enough in business to be able to discern a good idea from a bad, a wise investment from one that is going to take you down, and be able to read people enough to not be taken. Learn from your advisors, take classes, protect your brainchild from those who would capitalize on your ignorance and inexperience.

Creating a viable nanotechnology business, designed to make a profit, is the key objective in the planning and setup of the business. How the business is structured requires careful analysis of the number of owners and how to protect them via corporate structuring. Planning the dissolution of the business without tax encumbrances is as important as setting up the business.

Author Tim Berry created a business planning manual entitled *Hurdle: The Book on Business Planning*. In this book Berry wrote of a man who had spent 15 years trying to make his manufacturing business work and grew older and went further in debt from lack of planning. He quotes the man as saying, "If I can tell you only one thing, it is that you should never leave yourself without an exit. If you have no exit then you can never get out. Businesses sometimes fail, and you need to be able to walk away. I wasn't able to do that."

Berry cautions budding entrepreneurs that you should know what monies you need and understand that the money is at risk, cautioning not to bet money you can't afford to lose. This he notes is why the "U.S. government securities laws discourage getting business investments from people who aren't wealthy, sophisticated investors. They don't fully understand how much risk there is. Please, as you start your business, make sure you understand how easily money invested in a business can be lost."

Berry's advice is not to deter entrepreneurs, but to plan more effectively to prevent future heartaches and stress in business start-ups. Berry has his MBA from Stanford, is the president and founder of Palo Alto Software, and is founder of bplans.com, a software company that helps start-up businesses with professional all-in-one business start-up kits. While the software may not be specifically designed for a nanotechnology business, it can offer insights into what goes into creating a business plan, garnering funds, etc., of most general new businesses.

Creating the plan is key in developing the business proposals for investors, partners, or angels who will help to bring the idea to commercial fruition. Without a plan on who your customers are and their needs, a genius idea may only be beneficial to a limited audience and not a good commercial venture. Working through the business plan will help the owner to look at all facets of the business. It demands the time and due diligence to understand every detail of how to commercialize the nanotechnology idea. Years ago, when starting my first business, I was privileged to have a small business counselor tell me, "If more people did a business plan, there would be a lot less new businesses, but there would be more that were successful."

That being said, creating the plan for nanotechnology requires the core group or scientist that wants to take it to investors to detail the information so that the investors can grasp the concept and add their own business savvy to commercialize it. Many first-time entrepreneurs or those who are not schooled in business keep their plan for the business in their head and wing it each step of the way. This becomes problematic when multiple people enter the picture if the vision is not articulated properly. This also allows for little all-around growth of the idea that can come from a business plan that allows for finance, marketing, management, and other business skills to be involved. The skills and genius needed to create the nanotechnology idea require a different thought process than those needed to create a business empire. This was painfully apparent in the dot-com bust, when the genius ideas died because of the lack of business structure to sustain the development, growth, and dissolution.

The business plan helps to understand your part in the business. Many times when an entrepreneur think of setting up a business he is enamored of the idea of being his own boss, playing golf at will, or the idea of being independent and not having someone breathing down his neck all the time (his current boss). Entrepreneurs envision themselves as a much more understanding boss. And yet, anyone who has owned a business understands that when you own a company or business, you have many bosses, i.e., the customer, the bank, fixed costs, employees wanting their check on payday, the mortgage company wanting the payment so you can stay open. As for being independent, we go back to Berry, who says, "Owning a business doesn't make you independent—making money makes you independent. As long as you need money, you can't be independent."

Key Questions

In creating the business plan, understand its multiple uses. Do you want to use it for internal planning to articulate your vision? Do you want to use it for banks? Do you want to use it to secure a venture capitalist or angels? Do you want it for the structure it will give your company? All of these questions merit planning, as each will indicate the detail needed. Banks and financial institutions often want more detail on personal net worth and the business's financial position; some want monthly projections, some want to look at collateral. Personal investors, on the other hand, want you to provide proof such as market data, management track successes, competitors and your competitive advantage, financial projections, etc. Again, the key in writing the business plan is knowing the audience.

Some of the questions a business will have to ask itself for the plan are:

1. What are you going to call the business? Do you know how to research availability, register, and protect your name?
2. Do you know the patenting laws and how to patent and copyright or trademark your product?
3. Do you know how to go about getting any licensing and permits you need?
4. Do you know how to obtain your tax ID from the Internal Revenue Service?
5. Do you know the financial basics of running a company? If not, do you have someone on staff to protect your interests?
6. What is the mission of the company?
7. What is the company history?
8. What are the key product features?
9. What is your target market?
10. What is your competitive advantage? Why would a customer choose you?
11. What is your basic marketing strategy?
12. Do you know your market? How many potential customers do you see?
13. Have you done a market analysis to see the opportunities?
14. Do you have inventory?
15. How do you manage inventory?
16. What about sales and sales people?
17. Do you have a sales and expense forecast?
18. Do you sell business to business or retail?

19. Do you sell on credit or do you have accounts receivable?
20. How is your cash flow?
21. Are you on an accrual basis or cash basis for tax purposes?
22. Do you have an accountant or CPA?
23. Do you have an attorney?
24. Do you have insurance professionals for workers' comp, etc.?
25. Who are your advisors?
26. Who is your banker?
27. What are your start-up costs? Are they realistic?
28. What are you including in your start-up costs—office equipment, signs, building expenses, websites, product development, packaging, setting up retail, or supply chain?
29. Do you plan on bootstrapping (starting without initial start-up capital)? Do you know the risks?
30. What are your patenting costs?
31. Can you find the technical people you need for your company?
32. How do you plan to implement this?
33. Are you writing your plan in stages or all at once—from history to financials, profit and loss, expense forecasts, etc.?

The following is an outline of a potential business plan:

THE BUSINESS PLAN

I Title Page

All contact and ownership information is included on the title page. Some entrepreneurs like to add a very brief business description, slogan, or mission statement.
a. Business name, address, telephone, e-mail, and website
b. Name of owner(s)

II Table of Contents

Include a list of all sections of the business plan and the appropriate page numbers. Graphs, diagrams and other visual representations should also be identified. Items included as exhibits at the end of the plan (example: owner resume) should be clearly identified so that the reader can reference them while reviewing the plan.

III Mission Statement

The mission statement should describe why your company exists in the market place. Some companies use this statement as a foundation for management decision-making and publicly display it on promotional literature and in the place of business. Many entrepreneurs find it useful to make the mission statement brief and general enough to allow potential growth of product lines and services. Consider the difference between describing yourself as a company in the "automobile" business, and a company in the "transportation business." The mission statement is usually not changed for five years or more and so it is important for it to adequately portray your firm's identity and philosophy.

   a. Description of company purpose
   b. Identification of those served

IV Executive Summary

An overview of the content of your business plan allows managers, strategic partners, investors, or lending agencies to quickly grasp your concept and business direction, so that as they read the pages that follow, they have a clear idea of your intentions. Because the plan encompasses so many activities, the reader could fail to extract the owner's view of the most important information. You will find many uses for this summary as you move forward to promote your company, network in the business community, and work with vendors of business products and services.

   a. Brief description of the company history
   b. Purpose of the plan
   c. Goals of the business
   d. Description of the products and services
   e. Customers
   f. Management team experience
   g. Amount required from lender*
   h. Other sources of funds/collateral*
   i. Method of repayment*

V Target Market/Customer Base

An error in the determination of your target market(s) will not only adversely affect all other sections of your business plan,

---

* Items marked with an asterisk are added to the business plans being used to secure financing.

it will most certainly increase your advertising and promotion expense. For some businesses it is the difference between success and failure. In this section of the plan describe the most likely customers for your product or service. Who are they? Where are they? When and why will they buy from you? To be thorough you must also describe the target market between you and the end user of your offerings. For example, if you are a manufacturer, you may need a retailer or distributor. Without the retailer or distributor purchasing your product, the end user will never have the opportunity to purchase. You may need promotional literature such as product and price sheets for this "middle" market and you may even need sales assistance. Overlooking this market could result in underestimated expense.

Often your entire market of purchasers can be divided into segments, or groups of purchasers with common needs. Segmenting your market allows you to define and describe buyers' needs and habits as completely as possible. Accurate information about the size of your market and expected market share helps you predict potential income.

a. Characteristics of the target market:
   – Demographic profile (age, income, sex, education)
   – Business customer (industry, size, purchaser)
   – Geographic parameters
b. Size of the market/expected market share
c. Market segmentation
d. Customer buying habits (seasonality, quantity, average expenditure)

VI Marketing Plan

The marketing plan describes all activities involved in selling. It sets annual sales goals and examines the competitors' products and services and how your offerings are unique. Marketing is not simply advertising and promotion activities. Although these communication elements are extremely important, they are ineffective if you have not chosen products and services wanted and needed by your potential customers. The marketing plan should include a complete description of all offerings. Names, colors, assortments and other details are important to customer choice. If you have multiple products for multiple target markets, this is the section where those distinctions must be made.

If you are tempted to dismiss competition, ask yourself how your potential customer currently solves the same problem your offerings are intended to solve. What are the customers' choices when spending their financial resources? It can be helpful to develop a matrix that lists all your major competitors, their products and services, prices, methods of promotion, and location. By incorporating your own marketing information on the matrix, you can identify your firm's strengths and weaknesses. Your marketing section includes customer service policies. Small businesses often have an opportunity to compete with larger firms by offering flexible, courteous, customer-centered services.

The pricing of your product must consider competition and customer expectations, but it must also consider all expenses. It is not uncommon for early-stage businesses to: (1) believe they can sell at the lowest price; (2) misunderstand the importance establishing price policies at levels other than the end user level; and (3) overlook the relationship between pricing and other elements of marketing.

The location element of business planning once focused on a physical business site, customer access to that site, and transportation (logistics) related to the site. With advancements in technology, both start-up and existing businesses must examine whether the location for interface with customers is a physical location, cyberspace, or both. A website can be used to simply promote a business and its offerings, or it can be the actual market place where sales are consummated. Website development, performance, delivery systems, and payment activities are now a necessary part of the marketing plan.

Few businesses exist without advertising expense. The choices of strategy and media are many, but the choice to eliminate advertising says the entrepreneur cannot afford to communicate with customers. A lack of communication is directly related to a lack of customer spending and a lack of customer spending critically impairs the business's survival. Since advertising and other elements of promotion are legitimate business expenses, they must be incorporated in the price of the products and services.

a. Sales goals
b. Description of all products and services
c. Direct and indirect competition
d. Pricing objectives/methods
   - Wholesale and retail
   - Discounts and special allowances
   - Seasonality in pricing
   - Credit terms

    e. Location
- Where products/services will be sold
- Website
- Analysis of advantages/disadvantages
- Plant/store atmosphere
- Transportation

    f. Promotion activities
- Advertising
- Public relations
- Publicity
- Trade or business shows
- E-Commerce

    g. Packaging
    h. Customer service policies
    i. Sales training, management, and methods
    j. Growth strategies

VII Production and Operations Plan

A lack of production and operations planning causes entrepreneurs to underestimate start-up, maintenance, and growth expenses. The decisions in this section of the plan consider the "physical" health of the business. If the business is started at home, the entrepreneur should set criteria such as income, number of employees, or product expansion that will necessitate moving to a business site. Decisions made in this section affect the extent of company indebtedness, as well as the collateral of the business when it seeks out loans or investments.

    a. Facility
- Lease or purchase
- Size and floor plan
- Zoning, local regulations, taxes
- Renovation/expansion plans

    b. Equipment
- Machines/tools owned/needed
- Lease or purchase
- Maintenance procedures and costs
- Vehicles
- Telecommunications and data

    c. Production process and costs
    d. Suppliers/credit terms
    e. Transportation and shipping access and equipment
    f. Scheduling for completion of research and development

VIII Insurance
By definition, entrepreneurs are risk takers. They launch a new enterprise in a competitive environment with less than adequate capital and work more hours in the day than their corporate employee counterparts. Once the decision has been made to become an entrepreneur, risk management becomes a part of the job description. As a firm grows, the wise entrepreneur develops a risk management program with advice from an attorney, accountant, and insurance agent. Young firms are vulnerable and protection comes from evaluating and prioritizing risks and insuring against them. You can start by making a list of the perils your business faces. Identify which are most catastrophic, such as loss of life, damage to property, employee or customer injury resulting from a faulty piece of equipment or product. Take action to protect your business against these catastrophes first. Risks differ related to your industry and specific offerings, and gaps in coverage can occur as the business grows. Your risk management program should be evaluated annually.

a. Product liability
b. Personal/business liability
c. Business interruption
d. Vehicle
e. Disability
f. Workers' compensation
g. Unemployment
h. Fire
i. Theft

IX Management and Human Resources Plan
The people in any business are an important and expensive resource. Before developing this section of the plan, the entrepreneur must identify how the business will grow and what skills will be needed for that growth. If additional locations are planned, new managers will need to be hired or trained. If growth comes from development of new products, researchers and engineers may be needed. If growth will result from selling intensively to a small number of clients who buy on multiple occasions, employees that are capable of developing good relationships and delivering excellent customer service are needed. The obvious expense of human resources is salary and benefits. Less obvious is the cost of recruitment, selection, and training when turnover occurs. This section requires knowledge of state and federal regulations governing employer and employee relationships.

a. Key managers
   – responsibilities

- training
- reporting procedures

b. Personnel
   - number of full- and part-time employees
   - special skills/education required/continuing education
   - job descriptions and evaluation methods
   - benefits
   - wages, commissions, bonus plans
   - use of subcontracted personnel
   - policies
c. Organizational chart
d. Lists of stock holders and board members
e. Amount of authorized stock and issued stock
f. Professional assistance (attorney, accountant, banker, insurance representative, etc.)

X Financial Plan

Books and software packages can be purchased with formatted worksheets to produce the documents you need for your financial plan. The numbers used for each expense should be as accurate as possible based on current research. Identify any fluctuations that can be predicted such as increases in raw materials, lease or utilities in year two or three of your business. Estimate the month and year when additional employees will be hired and what they will be paid. A break-even analysis helps you understand at what point the business becomes profitable and allows you to set goals realistically. Without a financial plan you will find it nearly impossible to interest lenders or investors in helping you start and grow, because you have no facts to back up your enthusiasm and commitment to your venture.

a. Start-up costs (all one-time expenses such as equipment, deposits, fees, etc.)
b. Monthly expenses (ongoing expenses for lease, insurance, utilities, etc.)
c. Sources and uses of funds*
d. Balance sheets (opening day and projected three years)
e. Projected cash flow (monthly first year, quarterly year two and three)
f. Profit and loss forecast or statement (annual for three years)
g. Break-even analysis
h. Existing business (historical statements for three years*)

---

* Items marked with an asterisk are added to the business plans being used to secure financing.

    i. Personal financial statement of owner(s)*
    j. Assumptions used in preparation of financial projections

XI Attached Exhibits
    a. Managers' resumes
    b. Advertisements, news articles and other promotional documents
    c. Contracts, leases, and filing documents (Fictitious Name, Employer Identification Number, Articles of Incorporation)
    d. Letters of support
    e. Pictures of the product or service
    f. Marketing research
    g. Patents, trademarks, copyrights, license agreements
    h. Income tax returns (three years)*
    i. Invoices or estimates for facility or equipment purchases*

Pennsylvania Entrepreneur's Guide 2009, http://www.fcadc.com/incentives/pdf/Entrepreneur_Guide.pdf.

---

* Items marked with an asterisk are added to the business plans being used to secure financing.

This chapter will feature the structure of setting up the business, creating the business strategy, as well as deciding on the structure and partnering required to build a sustainable commercial scientific entity.

Deciding on the structure, whether a sole proprietorship, a partnership, or corporation, will require owners to decide how much time they want to contribute or their participation and at what level. It also requires a look at the initial costs, how they are going to finance the business, and what the tax implications and credits will be with each structure. For example:

Sole proprietorship: In this type of structure, the individual owner owns all assets and is recipient of all income, but he or she is also totally responsible for all liabilities and losses and creditors can go after his or her personal assets. Sole proprietorship requires a simple business name registration. This type of ownership does not create a separate legal entity or trade name. If you are not using your own name, you can register the company under a fictitious business name or DBA (doing business As). In the United States, it is dependent on the state you register in, and you can normally set this up through the county government with a small registration fee and the required newspaper advertisement posting for less than $100. Taxes go through your personal taxes, and the business income is normally shown on the Schedule C of the tax return.

Partnership: In this type of structure, all personal assets are at risk, as well as being responsible for both the partner's and employee's actions. A partnership requires a partnership agreement, deciding the terms, financial contribution, and outlining the contribution of each partner. It requires a simple partnership name registration. In the United States a uniform partnership act sets specific partnership agreements as the legal core of the partnership; thus the details can vary broadly, such as general and limited partnerships that define different levels of risk. Liquidation or termination of the partnership needs to be spelled out with buy and sell arrangements for the partners to be defined. Partnership also looks at the citizenship and residency requirements. Since the terrorist attacks of 9/11, a number of labs have become unavailable to those who do not hold citizenship, thereby limiting the partnering of foreign nationals to be fully involved.

Limited liability company (LLC): A corporation must have four characteristics: limited liability, continuity of life, free transferability of ownership interest, and centralized management. A LLC is structured similar to an S-corp with a combination of limitation on legal liability and the favorable tax treatment for profits and transferability of ownership interest.

Corporation: This is a government structure that is set up for the benefit of the shareholders to protect the assets that are invested in the company. A corporation is a separate legal entity that can own assets and incur liability. The structure provides that the liability is limited to the money and assets invested in the company as opposed to the personal liability in the structures mentioned above (Figure 3.1).

Using a standard C or the small business S corporation defines the liability structure of the corporation. The C-corp is structured to provide the best shielding from personal liability and the best nontax benefits to the owners. As a separate legitimate entity from the owners that pays its own taxes, this is the best for those companies wanting to raise major investment capital and going public. The business S-corp is often used for smaller ownership groups or family companies. The main difference with the S-corp is that the profits and losses go straight to the corporate owners without being taxed first, allowing the owners to take profits first before paying the corporation's separate tax on the profits, allowing the profits to be only taxed once, but twice for the C-corp. The C-corp is often chosen because of the personal liability shielding and the goals for growth. Corporations can switch from C to S and back, but the IRS has strict rules for this.

FIGURE 3.1
Corporation. A corporation is a separate legal entity that can own assets and incur liability. The structure provides that the liability is limited to the money and assets invested in the company as opposed to the personal liability.

In choosing the structure, the following questions must be addressed:

1. What is the tax year end?
2. Who signs checks, notes, loans, and contracts and what are the signing limits?
3. How are meetings held?
4. Voting issues—on the occasion that there is a voting tie, who gets the casting vote?
5. Changes in the business—how is the decision making structured?
6. Contracts—who decides if business enters into contracts, partnerships, or joint ventures?
7. How are shares valued in the event of a buy-out?
8. When can a shareholder be forced to sell or buy shares?
9. How will profits be split?
10. How are dividends declared?

Strategy

Answering a few key questions is critical in knowing where you are going with your product. First-mover advantages and ease of entry into your target market are key in setting the bar in nanocompetition. Mapping out your business strategy will define the commercialization of your product.

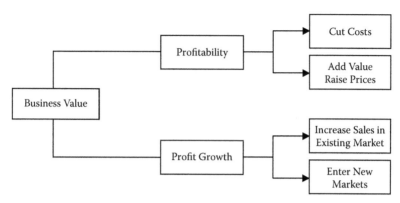

Once established, the company can increase profitability and profit growth.

FIGURE 3.2
This model shows profitability and profit growth decisions for businesses during the growth and development phases.

Strategy is defined as the actions that management takes to attain the goals of the firm. Defining those goals is most often done by the presiding board of directors the company chooses. The goals are set with the profitability or the rate of return the company will make on the invested capital and the profit growth or increase of the net profits over a period of time. A 10 to 20% yearly growth is an attractive choice for many companies. The advent of nanotechnology is creating new horizons in growth.

Nanogate, our case study for this chapter, is a leading international nanotechnology enabler located in Göttelborn, Germany. The company began in 1999 and has been regarded as a trailblazer in the nanotechnology field. Nanogate enables the programming and integration of additional properties—such as nonstick, antibacterial, anticorrosive, and ultra-low friction—into materials and surfaces. The company increased group sales by 38.6% to EUR 5.67 million in the first half of 2010 (previous year: EUR 4.09 million).

Understanding the potential for profit growth with a nanotechnology enabler or nanoproduct will require some research on the part of the business owner as to the product and industry he or she is developing. This industry knowledge will give the corporate governing bodies the criteria for setting the profitability and profit growth standards for their shareholders (Figure 3.2).

Early Mover Advantage

Becoming the first significant company to enter into the market offers what is called first-mover advantage (FMA). First-mover advantage for a company is the advantage that it has gained by being the first in the market segment, often allowing it to control the resources that subsequent followers may not be able to match. First movers, while taking the chance with the risks, frequently are rewarded with large profit margins and often move to a monopoly-like status. In nanotechnology, many of the first movers secured multiple patenting around their discoveries that can somewhat limit direction of growth or define partnering. Patenting will be discussed in depth later in this text.

Strategy to acquire this position of a first mover requires doing the due diligence and testing to see that the rewards outweigh the underlying risks; this is critical in the development and marketing of the product, as often a second-mover advantage company can capitalize on the initial surge of interest and resources if the first mover is not able to capitalize on the advantage. If the first mover cannot capitalize, it opens the door for other companies to compete more effectively than the first user, or earlier entrants. Thus the strategy of setting up the company is key to becoming a first mover and controlling the resources.

Setting strategy requires looking at the big picture to understand the value chain of the company. People often see the companies as a value chain that is composed of distinct value creation activities such as production, marketing materials management, R&D, and human resources. The production of the nanoproducts passes through all activities of the chain in the order set up by the company, allowing each activity to add expertise to the value of the product. Using the value chain of activities gives the products more added value than the sum of added values of all activities.

Identifying the areas and activities where the company adds value for the customers focuses on the customer-oriented activities, whether they be health needs, computer needs, transportation and gas needs, etc. The product is only as good as the need it fills for the customer.

In setting up the criteria for profitability and profit growth, the company needs to look at its target market to determine the value creation. The value creation is the difference between the V or price that the company can charge for the nanoproduct within the current competitive pressures and C or cost of producing the product. The value is measured by the difference between what a company can charge for the product within the current competitive environment and the costs that are incurred producing the product.

As far as it's about strengths and weaknesses, this model will help organizations identify those areas where they are adding value to the customer (strength areas) and those areas where they need attention to add values because value chain is all about how you do something extra for your customers that your competitors can't or don't.

When cutting costs or lowering the price isn't an option to increase profits, companies focus on increasing the attraction of the product to create the differentiation strategy. By making the product more efficient, visually or sensory appealing, or easier to use, they can differentiate their product from the competitor's and gain market share.

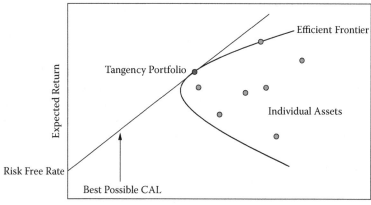

FIGURE 3.3
A tangency portfolio asks how much you would be prepared to lose in a worst-case scenario without bailing out of the market. The tangency portfolio shows the optimal combination of risks assets that maximize at each level of risk.

After defining the costs and differentiation, the company needs to pick a point on the "efficiency frontier" where there is adequate demand for the product, and set the price, configuring the internal operations to support the decision. The customer has to see added value over the competitor before it will pay more for the product (Figure 3.3).

Configuring the internal operations to support the strategy begins with the company operations being viewed as a value chain. A value chain is composed of a series of distinct value creation activities that include production, marketing, materials management, R&D, human resources, information systems, and the company infrastructure.

These activities fall under two categories: (1) primary activities that include creating the product, marketing and delivering it to buyers, and providing support, and (2) follow-up sales and customer service to the customers.

The primary activities include R&D, production, marketing and sales, and customer service.

Support activities are the underpinnings for the primary activities, allowing them to occur. They include information systems to manage inventory and track sales, logistics, and human resources. The support activities are the informational systems that manage and track the products and sales, logistics, and the human resources that cover all areas that keep your new business running.

As companies progress in the development of the product, they often find that it is more efficient and effective to expand their market in both production aspects and sales. In today's world, a finished product is often comprised through the efforts of going where the technology and labor are the most reasonable. By dispersing the value creation activities to the areas where they can be performed most effectively and efficiently, the company not only saves money, but picks up the intellectual capacity and cultural work secrets of various locations that will help its product strengthen in production. Utilizing the experience of the various locations, and the experience of the core team, can help the company to expand and realize greater cost economies by leveraging the skills of the foreign operations and transferring them within the company.

Because of the potential to manipulate the elements of the periodic chart, nanotechnology has the potential, through such areas as nanofabrication, to machine at an atomic scale and create new materials and systems. The materials that we were once familiar with will react much differently at the nanoscale. By leveraging the knowledge with the core competencies and skills, companies stand to learn more and develop higher-end products.

Choosing to become involved in a global strategic alliance offers several methods of expanding to the foreign markets that include licensing, franchising to host country firms, exporting, establishing joint ventures, or if the climate, workforce, knowledge, and political support are right, to set up a wholly owned subsidiary in a host market. Growth frequently entails acquiring established enterprises to give you the product differentiation and control in the market.

Global strategic alliances look at favorable markets as those that are politically stable, have free market systems, relatively low inflation rates, and low private sector debts.

The rapid evolution in the field of nanotechnology will also demand that the strategic analysis and planning be done on a more frequent basis than is done for normal companies. Moore's law states that technology will double in speed every 18 months and prices will fall. Just as the technology is moving that fast, the need for education of nanotechnology workers will increase as well, making the global option more of a reality to find necessary employees. Studies show that nanotechnology employee estimates will be approximately 2 million around the world by 2015. The breakdown of workers needed is: United States, 0.8–0.9 million; Japan, 0.5–0.6 million; Europe, 0.3–0.4 million; Asia Pacific, 0.2 million; other regions, 0.1 million.

This surge in nanoproducts has the potential to create 5 million additional nano-related jobs in the global market by 2015 that could pull us from the current economic struggles, yet increase additional educational and training pressures to prepare the workforce.

The ease of doing business index averages the country's percentile rankings on 9 different aspects of the business environment:

1. Starting a business
2. Dealing with construction permits
3. Registering property
4. Getting credit
5. Protecting investors
6. Paying taxes
7. Trading across borders

8. Enforcing contracts
9. Closing a business

The ease of doing business index ranks economies on their ease of doing business on a scale of 1 to 183. A high ranking on the index means the regulatory environment is more conducive to the starting and operation of a local firm. This index averages the country's percentile rankings on 9 topics, made up of a variety of indicators, giving equal weight to each topic. The rankings for all economies are benchmarked to June 2010.

Economy	Ease of Doing Business Rank	Starting a Business	Dealing with Construction Permits	Registering Property	Getting Credit	Protecting Investors	Paying Taxes	Trading across Borders	Enforcing Contracts	Closing a Business
United States	5	9	27	12	6	5	62	20	8	14
Singapore	1	4	2	15	6	2	4	1	13	2
Hong Kong SAR, China	2	6	1	56	2	3	3	2	2	15
Mexico	35	67	22	105	46	44	107	58	81	23
China	79	151	181	38	65	93	114	50	15	68
Germany	22	88	18	67	15	93	88	14	6	35
United Kingdom	4	17	16	22	2	10	16	15	23	7
Japan	18	98	44	59	15	16	112	24	19	1
Russian Federation	123	108	182	51	89	931	105	162	18	103
Canada	7	3	29	37	32	5	10	41	58	3

Economy Ease of Doing Business Rank—Top 10 Countries

1.	Singapore	142156241132
2.	Hong Kong SAR, China	261562332215
3.	New Zealand	3153212628916
4.	United Kingdom	41716222101615237
5.	United States	592712656220814
6.	Denmark	62710301528135305
7.	Canada	7329373251041583
8.	Norway	833658462018944
9.	Ireland	9113878155723379
10.	Australia	102633565948291612

Source: http://www.doingbusiness.org/rankings.

Analyzing the company resources and capabilities is reflected in a VRIO framework. The resource-based view focuses on (V) value, (R) rarity, (I) imitability, and (O) organizational aspects of the resources and capabilities of the company. Looking at a company from the VRIO perspective offers a competitive advantage.

Value: Only value-added resources can lead to competitive advantage, while non-value-adding capabilities can lead to a disadvantage. An example of this is a case study of IBM. IBM found itself being phased out in the early 1990s because its historical expertise in hardware was no longer competitive. The company had begun in 1930 making tabulating machines. In 1960 it became known for its mainframe computers, and in 1980 the PC. Seeing the competition writing on the wall, and with the ideas of a new CEO, IBM entered the more lucrative field of software and services, adding new value-adding capabilities. It sold its PC division in 2004 to Lenovo in China.

Nanotechnology companies need to constantly be looking at the value-adding capabilities of their products in this competitive science with many overlapping patents. With the amount of capital invested in the technology to bring it to the commercialization stage, it is key to keep an eye on the value for a return on investment. Just as IBM was the leader for many years, it would have had a slow death in the field of hardware competition.

Rarity: Knowing how rare the valuable resources and capabilities are to stay competitive. Having a valuable, but common resource and capability will give a company competitive rarity but not the advantage in the market. There are various marketing strategies, and a company has to look at the big picture in regards to return on investment (ROI). Saturating the market with a nanoproduct will undermine the novelty value or rarity, whereas marketing it as a valuable and rare product will give it the potential to have a modicum of at least a temporary competitive advantage. "If everyone has it, you can't make money on it." If it is no longer rare, it provides no advantage.

Taking IBM again as an example, the company earns $1 billion a year from its IP portfolio. Microsoft now files over 3,000 patents a year, as opposed to 5 filed in 1990. Both companies are making big money on the rarity factor. Nanotechnology companies need to closely monitor their intellectual property and continue to file patents as they grow, adding additional revenue and the rarity factor to their companies.

Imitability: Although intellectual property is covered in the patenting, copyright, and trademarking, products and licensing can still be violated. Blatant patent infringement is illegal, but reverse engineering, or inventing around the patent by changing a few things, is legal.

Thus the valuable and rare resources and capabilities can be a source of competitive advantage only if they are hard to imitate by the competitors. Nanotechnology is going to be a boom industry, money flowing like water, if we look at it by the billions of dollars that governments are investing in it. The race for riches will include those competitors interested in imitating the ideas for their own profits.

Tangible resources can be recreated through reverse engineering, while intangible capabilities such as the tacit knowledge, managerial talents, and motivation cannot. Looking at a competing firm's successful performance can make it difficult to imitate because of the casual ambiguity, or the means of putting one's finger on the actual cause of success in a company. A company needs a secret recipe, so to speak, of unique practices it has developed and used to make it or its production unique. Nanotechnology companies of the future will need to find their unique competitive advantage, or what makes their company special. Only valuable, rare, and hard-to-imitate resources can lead to sustained competitive advantage.

Organizational: We all need systems in place to make our lives run easier, and managing a business is no different. Nanotechnology companies will have to look at their organizational structures to develop and leverage the full potential of its resources and capabilities. Have you ever watched the credits at the end of a truly great movie, astounded at the number and quality of people, jobs, songs, and resources it took to make a stellar picture? This is an example of what is called complementary assets—the combination of resources, individual assets, organizational attributes, smiles, laughter, hard

TABLE 3.1

Ease of Doing Business: Top 10 and Bottom 10 Countries

The World Bank ranks the 183 economies of the world on the ease of doing business. High scores signify favorable business environments, showing a regulatory environment that is conducive to the operation of business, while low scores signify unfavorable business environments and indicate a regulatory environment that is not favorable to the business.

Country	Ranking	Country	Ranking
Singapore	1	Niger	174
New Zealand	2	Eritrea	175
Hong Kong, China	3	Burundi	176
United States	4	Venezuela	177
United Kingdom	5	Chad	178
Denmark	6	Republic of Congo	179
Ireland	7	Sao Tome and Principe	180
Canada	8	Guinea-Bissau	181
Australia	9	Democratic Republic of Congo	182
Norway	10	Central African Republic	183

Source: World Bank, Doing Business 2010, Ease of Doing Business Rankings on the 183 countries, http://www.doingbusiness.org/economyrankings/.

work, dedication, attention to detail, and clarity of vision in a social complexity, or the interdependent ways that the firm is typically organized.

These organizationally embedded capabilities that make up the uniqueness of the firm are what is hard for rivals or competitors to imitate. The uniqueness of the social capital that makes up the company often gives it its source of competitive advantage. Thus creating the team is critical for the success of the venture and what often gives it the sustainable competitive advantage.

The VRIO framework, if revisited regularly, can give the company a long-standing competitive advantage in the marketplace and value to investors.

CASE STUDY: FIRST MOVER ADVANTAGE

PRESS RELEASE

Nanogate formally closes deal for takeover of Eurogard B.V.

Takeover strengthens market position and opens up further growth potential—transaction successfully concluded Göttelborn, Germany, 8 June 2011. Nanogate AG (ISIN DE000A0JKHC9), the leading international integrated systems provider for nanosurfaces, has successfully closed the deal to take over the Dutch company Eurogard B.V. Nanogate and Eurogard aim above all to achieve growth together by expanding international business further, developing new applications and acquiring new customers. Nanogate expects to increase Group sales in the 2011 financial year to more than EUR 30 million and record an EBITDA margin of at least 10%.

Eurogard specializes in enhancing surfaces on two-dimensional components and is the global market leader in the lucrative specialist sector of coating transparent plastics. In the 2010 financial year, the company's sales were in the high single-digit millions and its EBITDA yield was in double figures. Eurogard is free from debt and generates a positive free cash flow. Its operating business is concentrated on the buildings/interiors, aviation, and automotive/mechanical engineering sectors. Eurogard coatings can be used on aircraft windows, utility vehicles, building elements and ski goggles, for example. Nanogate expects the new company to be fully integrated in the third quarter of 2011.

Ralf Zastrau, CEO of Nanogate AG, commented: "Nanogate's equity holdings in Eurogard B.V. and GfO AG put the company in an excellent strategic position. We offer a unique and complete technology portfolio with a broad range of applications. In addition to this, Nanogate covers the entire value chain like no other company—from materials expertise to process integration, right through to mass production. In operational and strategic terms, Nanogate is well positioned to grow faster than the market for surface coatings. Our aim is to significantly expand our market share."

Nanogate on Twitter: http://twitter.com/nanogate_ag

If you have any queries, please contact:

Christian Dose (financial press and investors)
Cortent Kommunikation AG
Tel. +49 (0)69 5770 300-0
nanogate@cortent.de

Liane Stieler-Joachim
Nanogate AG
Tel. +49 (0)6825 9591 220
liane.stieler-joachim@nanogate.com

Nanogate AG
Zum Schacht 3
66287 Göttelborn, Germany
www.nanogate.com

NANOGATE AG:

Nanogate is the leading international integrated systems provider for nanosurfaces, concentrating primarily on enhancing high-performance surfaces. The firm, which is based in Göttelborn (Saarland), enables the programming and integration of additional properties—such as non-stick, antibacterial, anti-corrosive and ultra-low friction—into materials and surfaces. As an enabler, Nanogate gains a competitive edge for its customers by means of product refinement using chemical nanotechnology. Nanogate covers a wide range of industries, functions and substrates. The company thus provides a decisive interface for the commercial use of chemical nanotechnology and bridges the gap between the suppliers of raw materials and industrial conversion into products. In doing so, Nanogate concentrates as an enabler on one of the most attractive segments in the industry. Nanogate has a unique combination of extensive materials expertise paired with comprehensive, first-class process and production know-how. As a systems provider, Nanogate covers the entire value chain, from the purchase of raw materials, to the synthesis and formulation of the material systems, right through to the enhancement and production of the finished surfaces. Nanogate focuses primarily on plastic and metal coatings for all surface types (two- and three-dimensional components).

The Nanogate Group currently has approximately 250 employees in all and since commencing operations in 1999 has been a trailblazer in nanotechnology. The company has first-class customer references (e.g., Audi, BMW, Bosch-Siemens Haushaltsgeräte, Junkers, Kärcher, Hörmann Group, Opel and REWE International AG) and many years' experience of different industries and applications. Several hundred projects have already gone into mass production. Nanogate has also entered into strategic cooperations with international companies such as the GEA Group and Dow Corning. Nanogate consists of Nanogate Industrial Solutions GmbH, Eurogard B.V., FNP GmbH for products in the sport/leisure sector, majority stakes in Holmenkol AG and GfO Gesellschaft für Oberflächentechnik AG, and an equity holding in sarastro GmbH.

DISCLAIMER:

\-

Nanogate generates record sales in H1 2010 and anticipates positive consolidated net income for H2 2010

Göttelborn, Germany | Posted on September 30th, 2010

ABSTRACT:

Back on a growth path: sales rise by 38.6%, earnings up—Forecast for 2010: increase in sales to at least EUR 16 million, positive consolidated net income anticipated in second half—Outlook for 2011: sales set to reach upwards of EUR 25 million, company expected to return to full profitability—EBIT margin of 15% targeted for medium term—Majority stake in GfO to contribute towards growth

Göttelborn, Germany, 30 September 2010. Nanogate (ISIN DE000A0 JKHC9), a leading international nanotechnology enabler, returned to growth in the first half of 2010. The company's sales reached a new high in the first six months of the year. Earnings also improved alongside the sharp rise in sales, despite one of the Group's subsidiaries incurring considerable one-off expenses. In August Nanogate arranged a majority investment in GfO Gesellschaft für Oberflächentechnik mbH. With this move the company became one of the leading European system providers for high-performance industrial surfaces. Nanogate expects to post record sales of at least EUR 16 million for the full year 2010. Considerable profit growth will significantly boost the year-on-year result, leading the company to expect positive consolidated net income in the second half. In the 2011 financial year Nanogate intends to exceed EUR 25 million in sales and return to full profitability. In the medium term the Group is striving for an EBIT margin of 15%.

Ralf Zastrau, CEO of Nanogate AG, said: "Nanogate has successfully returned to a growth path. Operationally and strategically, we will attain a new dimension in 2010. Our new GfO equity holding will prove profitable before the year is out. Our extensive investments will also pay off this year. Nanogate will experience faster growth in sales and earnings in the medium term. For 2011 and 2012 we anticipate a clear increase with ambitious targets when the success from our growth strategy takes full effect."

Group sales up almost 40%

Nanogate increased Group sales by 38.6% to EUR 5.67 million in the first half of 2010 (previous year: EUR 4.09 million). The share of international business rose to 43.6% of total sales (previous year: 40.9 %). The overall performance increased by 23.1% to EUR 7.03 million (previous year: EUR 5.71 million). Earnings improved as expected in the first half of 2010. Consolidated EBIT came to EUR –1.7 million (previous year: EUR –2.0 million). This figure was diminished by one-off, non-recurring expenses for realignment and restructuring at the portfolio company Holmenkol. One-off, non-recurring expenses came to over EUR 0.6 million. Adjusted for these factors, consolidated EBIT accordingly came in at around EUR –1.0 million. Consolidated net income totaled EUR –1.15 million (previous year: EUR –1.25 million); adjusted for non-recurring expenses the figure was EUR –0.8 million.

Cash flow from operations amounted to EUR –1.15 million (previous year: EUR –1.16 million) but was also severely impacted by the one-off expenses resulting from the realignment and restructuring at Holmenkol. Strong sales and earnings growth should result in positive operating cash flow in the second half of the year. Cash flow from investing activities came to EUR –1.6 million (previous year: EUR –2.0 million). Investment activity will continue to be reduced in the second half of the year according to plan, excluding the GfO investment.

Significant growth anticipated in the second half

In the financial year 2010 Nanogate will improve sales and earnings considerably. The innovation offensive launched in 2009 and the ongoing opening up of markets will contribute to the record sales expected. Furthermore, the company is anticipating a boost to growth from the majority stake in GfO. GfO's business is currently growing as planned and the company's order book is well filled. On current estimates Nanogate will report record sales of at least EUR 16 million in the financial year 2010 (previous year: EUR 10.7 million).

Consolidated net income for Nanogate will also improve sharply over the course of the year—in comparison with both the first half of 2010 and the 2009 financial year. Nanogate is therefore anticipating a profit for the second half of the year and positive operating cash flow. However, the profit for the second half of the year will be unable to fully compensate for all the one-off expenses incurred in the first six months.

EBIT margin to reach 15% in the medium term

In the medium term Nanogate is reckoning with faster growth. The majority stake in GfO alone should increase the pace of growth by 10%. Additional sales and earnings are expected from the investment from the first quarter of 2011 onwards. In 2011 and 2012 sales should increase considerably, both organically and as a result of the GfO acquisition, with next year likely to exceed the EUR 25 million mark. Nanogate aims to return to profitability at Group level as early as 2011. In the medium term the Group is striving for a consolidated EBIT margin of at least 15%. Additional external growth is also part of Nanogate's expansion strategy.

Innovation-driven competitive advantage:

Using this guiding motto, Nanogate AG launched a comprehensive innovation offensive in 2009 in order to open up new growth prospects for its clients. To achieve this, Nanogate increased its investments and funding significantly—in particular in the 2009 financial year—and has since then presented a multitude of innovations. Since going public in 2006 the company has already invested several million Euros in developing new technology platforms and has transformed these into marketable products. Nanogate has successfully illustrated its expertise in more than 180 cases, using innovations to provide its clients with a competitive advantage.

4

Intellectual Property

The first step to protecting the intellectual property of a company is to use the U.S. Patent and Trademark Office (USPTO) (http://www.uspto.gov/). The USPTO is the source for all information needed on patents, trademarks, and IP law and policy. Patent systems vary around the world among industrialized nations, but most use the U.S. patent system as a model. As such, we will use the U.S. model to discuss this.

A patent, as ruled by the U.S. Supreme Court, is anything that is "a product of human ingenuity" and can be protected by patent. U.S. law divides the various interventions that can be patented into four broad categories:

1. A process, e.g., a particular way of combining chemicals to produce a medicine
2. A machine, e.g., a piece of diagnostic equipment
3. Something that can be manufactured, e.g., a computer chip
4. Composition of a matter, e.g., new pharmaceutical drug or a new plastic

The U.S. Patent and Trademark Office (USPTO) (http://www.uspto.gov/) oversees the patent process. Statistics show that the USPTO granted over 170,000 patents in 2005, which averages to a patent being granted every 3 minutes. The *Business Journal of Phoenix* reported that on Valentine's Day 2006 the USPTO issued "Patent No. 7 million to John O'Brien at DuPont Co for a cotton like, biodegradable polysaccharide fibers used in textile applications."

The patent system provides disclosure of information about inventions. By teaching the world's public how to make and how to use the invention in the best way that the inventor knows, the patent system will grant him or her an exclusive ownership of his or her innovation. Patents are frequently licensed to a third party who wants to use the invention. The USPTO requires that the technology patented be shared; the third party paying for using the patent will often pay a fee for licensing, and royalties on the sales. Infringements on the patent from those who use it without permission are open to lengthy lawsuits for patent infringement and payment for damages.

The patenting of nanotechnology is bringing its own set of complexities to the patenting realm because of the new tools like the scanning tunneling microscope (STM) and similar tools that affected the nanotechnology applications because they did not conform to the existing classifications of intellectual property (IP).

While it refers to creations of the mind, IP can be inventions, literary and artistic works, and symbols, names, images, and designs used in commerce. It is divided into two categories: industrial property, which includes inventions (patents), trademarks, industrial designs, and geographic indications of source; and copyright, which includes literary and artistic works such as novels, poems and plays, films, musical works, artistic works such as drawings, paintings, photographs, and sculptures, and architectural designs. In theory, patents can be licensed. In practice, the red tape in licensing the patent would strangle the industry.

Nanotechnology-centered IP is characteristically different because nanotechnology is often developed through multidisciplinary expertise in fields such as biology, chemistry, engineering, and materials science. This creates additional filing that requires input from a team comprised of scientists that can represent any/many of the disciplines while collaborating on projects with multiple elements, each of which might require multiple IP licenses.

The USPTO has been struggling to keep up with the savvy investors and complexities of the wave of new nanotechnology patents. It has identified over 253 separate international patent classes.

Bruce Stewart, CEO of Arrowhead Research Corporation, a company focused on acquiring and commercializing nanotechnologies in the biomedical sector, notes: "As more drugs come off patent and pharmaceutical companies recognize that they can solve problems such as solubility and toxicity with nanotechnology, the global market for nanoparticle-based therapeutics could explode."

As investors, angels, and large corporations look to the high influx of patents that are recognized as solving the problems of toxicology and solubility with nanotechnology, the global market for nanotechnology is about to explode. Public interest foundations and universities hold the most patents, with most from the fundamental sciences; the nod turns toward the universities, who are turning out 4,000 to 5,000 patents per year.

This is a twofold effect for universities who provide most of the nanoresearch labs and can get the patents done for their schools, but also revenue from patents is becoming a key part of the financial strategy of major universities. The push is on for more universities as the patent of nanotechnology is the first push toward the university as opposed to, in the past, the private companies that drove the trends in IP, which were more closely held to protect company interests.

University patents in the past often took the form of building block patents, which protect fundamental concepts on all subsequent patents. These early block patents are very lucrative—substantial licensing for the university holding them—as they propose that the basic fundamentals of nanotechnology are patented, creating an entirely different landscape for future nanotechnology IP.

Nanotechnology is a platform technology. As such, the technology can have applications in diverse products and industrial sectors. A carbon nanotube that is used to transport a substance can be used in cosmetics, medicine, and to build electronic circuits and devices from single atoms and molecules, such as tiny transistors and one-dimensional copper wire. Why?

For example, carbon is naturally abundant, the fourth most abundant element in the universe. It is a nonmetallic element that forms the basis of most living organisms and is critical to the health and stability of the planet via the carbon cycle—interconnecting all the organisms on our earth. The carbon molecule bonds with a wide range of other elements and can form thousands of compounds. Its changing structure, depending on how and with what it bonds, makes it a very unique and versatile element that is very useful in multiple industries.

Thus one might patent the nanotube, which would create far-reaching control on every other industry wanting to use the nanotube to transport medicines, chemicals, etc. The good news is that he or she sells the nanotubes and other companies purchase them and use them. Cross-licensing is used as a way to get past the patent infringement and litigation that companies may find themselves in if they cross the line.

Intellectual property rights are protected under various federal and state laws; without protection, the property falls into the public domain and may be used by any party without license. Intellectual property rights protect the commercial interests of the company at the various stages of design, manufacturing, and product operation. A key strategy as you begin this process of obtaining the safeguards needed is to systematically build a portfolio of different intellectual property rights that will protect the different areas of the company's technology and commercial interests while preparing an IP portfolio that can attract and obtain investors.

At the design and development stage, copyrights and trade secrets can be immediately enforced, as explored later in this chapter. Once a product or service is developed, issued patents and trademarks protect the technology and associated names and symbols. Enforcement of intellectual property rights can be achieved through licensing, litigation, and other business means; effective acquisition of intellectual property rights must be done legally. Patents and trademarks require applicant action and response within critical filing deadlines.

Patents are among the only protections from infringement by large corporations for small companies and nanotechnology start-ups. As companies start to grow, the ability to keep trade secrets decreases and patents become the chief method of intellectual property protection. It is important to keep in mind that a patent application gets published 18 months after filing, unless the applicant opts out, in which case a foreign patent may not be pursued for the invention. Note that unless the applicant opts out and forgoes foreign filing, the description of the invention will end up in the public domain and be accessible to competitors, whether a patent issues or not.

Because of the interdisciplinary nature of nanotechnology, this can create a special problem for inventors. The U.S. Patent and Trademark Office houses seven different technology centers, including the biotechnology and organic chemistry center, the chemical and materials engineering center, and various art units within each center, such as the metallurgy unit and the polymer chemistry unit within the chemical and materials engineering center, but none dedicated specifically at this point to nanotechnology. The lack of focused expertise combined with the understaffed state of the USPTO (which is discussed later) is likely to result in the improper rejection of patents due to a mistaken conclusion that the taught matter is not new or has overly broad patents, giving the owner excessive control over a particular area.

One such example was a recent flood of information technology (IT) patents that overwhelmed the USPTO. The heavy load and lack of information resulted in the one-click Amazon.com patent that was criticized as being too broad, monopolizing the field. The effect of challenging an overly broad patent held by a competitor is a costly process for both, while the company being challenged must spend much time and money defending its patent—time that could better be spent on company growth.

One such example of a patenting debacle that can happen is *Eastman Kodak v. Polaroid. Eastman Kodak v. Polaroid* was a case over the ownership of patents that took over 20 years to resolve. Seven patents upheld by Polaroid led to the total destruction of Kodak's instant photography business. Kodak had to pay over $3 billion in infringement damages, compensation, and legal fees.

The importance of protecting intellectual property rights in the field of nanotechnology cannot be overlooked, and the USPTO has reached out to the nanotechnology community for solutions. Recently the Foresight Institute and the USPTO held a meeting to address the nanotechnology patent issues, resulting in "a set of Nanotechnology specialists within the USPTO and in communication with each other could unify prior art searched and ensure more accurate consideration of Nanotechnology patents and increased quality of granted patents" (Bastani and Fernandez, 2001).

The current patent process is a two-edged sword, explaining much of the reasoning nanotechnology companies have for taking the route they do. On one hand, getting one broad patent is risky; on the other, obtaining each necessary patent is very costly. Some nanocompanies have chosen to take a risk and obtain one broad patent for the field in which they are trying to market in. Their reasoning is that it is too expensive to obtain multiple patents, and they hope that the competition will be more likely to spend their time and money on new technology innovations rather than on challenging competing patents.

Conservative, less risky people decide to gain multiple patents ensuring their rights to that particular field. This has proven to be a surprisingly risky choice as well, as they are spending a large percentage of their funding on multiple patents and do not have enough for research and development. Although experts believe that nanotechnology will eventually be very lucrative for the inventor and the investor, there is still no hard evidence on the mass scale for investors. Risk can equal great reward, and in the case of nanotechnology, the risk of patenting is projected to equal much greater rewards.

The nanotechnology patent industry is competitive with numerous patents being filed every day. Patents can strengthen a researcher's reputation and enhance his or her résumé in the scientific community. In the business world, patents can create barriers to entry, increasing return for first movers. Savvy patenting can help companies occasionally shut down existing companies, as well as potentially deter newer companies from entering the marketplace. Patents have become an essential component to business enterprises, enabling the generation of revenue through licensing, and increasing their value, as evidenced by recent stock fluctuations in response to patent issuances and court decisions, as referenced by Serrato, Herman, and Douglas (2005) in "The Nanotech Intellectual Property Landscape."

In nanotechnology, this can be frustrated by the fact that patents are often sent to different centers at the Patent and Trademark Office for review. For example, different examiners in centers reviewing nanotube patents may be reviewing different prior art, and this may result in the issuance of patents that would have otherwise been rejected (Serrato, Herman, and Douglas, 2005).

As claims determine the scope of the patent, the recent patent trend "may simply be another manifestation of a culture consumed by property rights. There is infringement of a claim that takes place when there is a literal infringement or infringement under the doctrine of equivalents. It is possible that a single patent with a broad claim might be infringed by a large number of different companies developing the technology, while fifteen patents with limited claims might not be infringed by any companies in the field. The outlook of the intellectual property future is unknown" (Serrato, Herman, and Douglas, 2005).

A recent report proves the last article to be true. The IP outlook has been unknown for the entirety of the nanotechnology field's history. According to the report, as of March 2010, 6,000 patents for nanotechnologies had been awarded by the USPTO alone. Many of these patents were granted under the very broad field of science. The U.S. government was the largest patent patron for 2008. With research and development funded by the Department of Energy, Air Force, National Institutes of Health, Army, Department of Defense, Food and Drug Administration, and National Cancer Institute, one would assume they would spend more time detailing these patents. The absence of labeling rules on products containing nanoparticles makes it difficult to determine the extent of their use. Also, it makes it hard to understand the economic influence it has on our environment and society (IP Watch).

Patenting has its own game. The increase of patents being filed and issued is often an aggressive IP tactic called patent flooding. These "killer category patents" create conflict and limit operations in some companies, thus pushing the need for the cross-licensing. Cross-licensing is often the answer, but not always.

Aggressive companies issue multiple incremental patents that can surround a company of interest's IP, causing a deadlock where neither company can use the licensing without infringing on each other's IP, forcing them to cross-license to each other, giving the aggressive company access to the company of interest's IP. This idea of patent flooding is causing companies to file multiple or layered IP patenting for more protection. Instead of filing just one patent, informed companies may frequently file multiple patents in the following layers: (1) composition of matter patents, (2) process patents, and (3) application patents.

Another possible answer is patent pools that could further commercialization. A patent pool is a group of patent holders or a consortium of at least two companies that agree to cross-license their patents relating to a particular technology and is then collectively managed. A patent pool is defined as "the aggregation of intellectual property rights which are the subject of cross-licensing, whether they are transferred directly by patentee to licensee or through some medium, such as a joint venture, set up specifically to administer the patent pool" (see Joel I. Klein, "An Address to the American Intellectual Property Law Association, on the Subject of Cross-Licensing and Antitrust Law (May 2, 1997)," noting that *United States v. Line Materials*, 333 U.S. 287, 313 n.24 (1948) states that the term *patent pool* is not a term of art).

While patent pools were notorious in the past, a particular group in the 1930s and 1940s filed numerous patents, controlling 94% of the industry, allowing them to set prices unreasonably high on their products. These actions violated in the antitrust laws and the Supreme Court dissolved the pool. Today the U.S. Department of Justice and the U.S. Federal Trade Commission have recognized the importance and created Antitrust Guidelines for Licensing of Intellectual Property to help the survival of many companies in this era of rapid technological innovations.

While many countries do follow the USPTO, other emerging models are showing U.S. flaws with regards to nanotechnology, the most important being the turnaround time of sometimes 24 to 36 months. Even the "fast track" patent process is not felt to be sufficient for the rapid nanotechnology development. The 24- to 36-month timeline threatens to make some patents obsolete before they are even issued. This time delay is critical for companies looking for Venture capitalists who typically want assurance of the patented before they will invest in a company. Venture capitalists need to have strong company portfolios to remain competitive.

Although it takes additional time, money, energy, etc., some start-up companies, to circumvent the patent delay, often go for licensing the rights to the patent that is pending so that they can move forward to get the funding they need to continue. The patent rights and licensing is the starting point for companies to begin attracting their seed money from the investors.

Quantum Insight (featured in Case Study 3) is located in Menlo, California, and has had exposure to hundreds of nanotech start-up companies, watching the tactics and strategies from venture capital (VC) firms, corporate VC, and start-ups in the nanotechnology arena. It explains the IP licensing model as follows. "The IP licensing model has the advantages that it allows the nanotech start-up company to avoid the expense of setting up manufacturing and sales channels—both expensive propositions. The way the IP licensing model works is that a company develops IP, then licenses it to other companies for commercial applications, and finally collects a royalty on the use of the IP. The royalty revenue is then used to fund more IP creations." For more information visit http://www.quantuminsight.com/papers/030915_commercialization.pdf.

Quantam Insight officers Anthony Waitz and Wasiq Bokhari created a report entitled "Nanotechnology Commercialism Best Practices." The paper, written from observations of successful and unsuccessful tactics and strategies of several hundred nanotechnology start-up companies, offers a focus on the key factors for nanotechnology commercialism success.

"From our point of view, inception of a company is synonymous with the acquisition of the company's initial Intellectual Property (IP)." Because of the process involved in securing the correct patents and license, many start-up nanotechnology companies obtain their initial IP from government or university labs, with the filer of the patent involved in the commercialization of the technology. "Most commercialization efforts start with taking steps to protect IP through the filing of patents."

Yet, patenting a nanotechnology prototype or idea can be a grey area. It is hard to figure out the details of what to patent because nanotechnology is so broad. Many times, a company, whether out of economics or lack of understanding of the process, may patent just the initial idea or a small portion of what it should have patented. Other companies will then patent the other cloudy areas of the patent that the creator of the idea should have patented, and in effect hold the company hostage to do business with them as they hold key patents for potential development of the product. Generally, the first to patent will have the best chance of winning the broadest patents. Like any "hot" technology area, says Michael Masnick in an article in Techdirt,

> it does not take long for massive, innovation hindering patent thickets to spring up. It effectively makes it impossible to bring anything to market unless the company obtains a large patent portfolio or has extremely deep pockets. A new report is suggesting that the latest "hot" area to get patent crazy is Nanotechnology. That same report has suggested that the biggest patent patron in Nanotechnology is the federal government. The federal government spends citizen's tax dollars to lock up many of these new inventions. No one has yet been able to credibly explain why federally funded research should get a patent. In the past, it was determined that federal documents could not be covered by copyright for this reason, but why doesn't that extend to patents? This unanswered question is especially disturbing because it seems clear what will eventually happen. The nation's tax dollars pay for the research, and then that same research is transferred over to a private company who uses the monopoly rights to keep the product expensive and limit further innovation.

Masnick's article, entitled "Next Tech Area to Be Hindered by Patents: Nanotech ... and Much of it Is Funded with Your Tax Dollars," looks at the implication of using tax dollars to fund nanotechnology.

Michael Berger wrote of this patent rush in an article entitled "Legal Implications of the Nanotechnology Patent Land Rush," likening it to the Oklahoma land rush of 1889. "The 'Sooners' in this nanotechnology patent land rush may be the ones who were issued what some say are 'unduly broad' patents early on, in the hope of getting a windfall of nanotechnology intellectual property (IP) rights. While not violating the rules intentionally, upon re-examination, some broad patents might not hold up to the United States Patent Act's requirement for full and complete disclosure. This uncertainty provides fertile grounds for possible litigation over nanotechnology patent claims based on broad and imprecise definitions and descriptions." The "Sooners" (basically on a first-come, first-served basis) in the form of universities and companies, from start-ups to multinational conglomerates, are "rushing to aggressively stake out their turf in the nanotechnology patent area" (Berger).

"As scientists sort out and document the results of their research, corporate entities continue to seek and carve out far-reaching patent rights in what is now a full scale patent 'land grab,'" Dr. Raj Bawa tells Nanowerk. "As this trend unfolds, uncertainty is growing amongst researchers, developers, policy-makers and investors regarding who really owns what particular swath of technology in the rapidly-expanding body of nanotechnology intellectual property. Some fear that the far-reaching patent rights provided by early nanotechnology patents clearly overlap" (Berger).

Dr. Bawa is a registered patent agent and holds a faculty position at Rensselaer Polytechnic Institute in Troy, New York. He serves as advisor to the Office of Technology Commercialization at the institute. He spoke to Berger about the problems of nanotechnology patenting, noting that "some commentators, ranging from university experts to government agencies, put the blame on the common trend of uncertainty and patent overlaps on problems at the U.S. Patent and Trademark Office (USPTO)." With a technology growing as fast as nanotechnology, the USPTO is experiencing delays in implementing nanotechnology training for examiners. At the beginning of the nanotechnology boom, the patents that were rewarded were too broad to be considered today, contributing to the uncertainty of granting patents of questionable validity and scope. While all of this is being worked out and under review, there is "a growing backlog of unexamined patent applications and increasingly lengthy periods for patent pendency." The general consensus is that for those early movers who saw the gold mine of nanotechnology, and were awarded the broad patents, there needs to be compensation, because the lack of understanding of the patenting body that issued the broad patents was not of those who applied (Berger).

The U.S. Patent and Trademark Office has had a continuing effort to improve the ability to search and examine nanotechnology-related patent documents. The agency has also created over 250 cross-reference art collection subclasses for nanotechnology. The subject of the definitions used in patent applications has become a major issue in potential legal conflicts over patent claim coverage.

Locating a local small company in a rural Pennsylvania area that reviews patents, this writer was privileged to learn more about the patenting activities at this level. The company contracts from the government, employs entry-level workers with no higher educational criteria, and puts the workers through a vigorous 2-week training program. It pays a higher than average wage for the area, is selective in hiring, and pushes the confidentiality of the materials it is working on. When the training is complete, the company sets the workers out "on the floor" in cubicles to work through the mountains of patents, pushing the patent workers on a relatively high quota with a minimum of mistakes to keep their jobs. Turnover at this specific company remains high. This is just one small company in a small town. Standardization, as will be covered later in this text, will need to incorporate the definitions of terms used.

In nanotechnology terminology there is considerable confusion over the language used. Surprisingly, there is no clear and generally used definition for the term *nanotechnology* itself, not to mention such words as *nanoparticles*, *nanostructures*, or *nanomaterials*. While high-level university graduates study the language for years, the patenting agent's short training is held to a list of definitions that are often broad terms and even ambiguous descriptions of key words like *carbon nanotubes* or *quantum dots* in patent applications. Berger posits that this is "a sure recipe for conflicting terminology and a dispute over what was meant and intended." Bawa and his colleagues suggest that

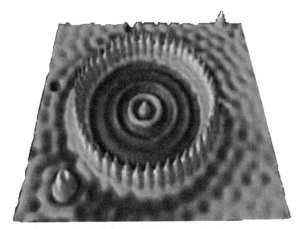

FIGURE 4.1
By using a nanoscale tip on scanning probe microscopes (SPMs) scientists can now see the atom. A closer look at this picture shows the individual atoms of the corral and the atoms that make up the underlying surface. (From M.F. Crommie, C.P. Lutz, and D.M. Eigler, *Science*, 262, 218–220, 1993.)

owners of nanotechnology patents that lack a clear definition of claim terms "may have to live with the uncertainty as to whether someone will dispute their claim at some point down the road" should they file patent applications that are not as broad in scope as a general strategy to avoid downstream problems in the future (Berger).

There is a direct correlation with the lack of early funding for nanotechnology and the issues around the validity of the patent laws as investors, interested in the new technology, are hesitant and do not want to get involved in an industry with so much grey area until it is more proven.

Both an offensive and a defensive strategy need to be used when obtaining intellectual property rights. Patents offer protection for processes that are novel, useful, and nonobvious, as well as the functional concepts, methods, and apparatus used in the prototype and production of the end products. Although it is costly to build an effective intellectual property portfolio, there will be a large drain on resources if there is a need to fight over patent disputes.

The Agreement in Trade-Related Aspects of Intellectual Property Rights (1994) offers criteria to help prevent patent disputes. This agreement defines patentable matter as "any invention that involves an innovative step and has a potential industrial application." The strategy in using a patent is to protect and advance the original idea through disclosure and teaching of the details of the invention to the public, and in exchange, the inventor or owner is rewarded the legal rights of ownership that are granted for a period of 17 to 20 years, depending on the date of the filing of the patent.

While it is very costly and lengthy to obtain a patent, cases such as *Kodak v. Polaroid* show the necessity of having your i's dotted and your t's crossed in the nanotechnology realm. Opting for a cheaper way to protect oneself needs further reflection, as different countries have different patent laws. In terms of patent obtainment in the nanotechnology field, there are several other countries with more advanced patenting than the United States. In Europe, Japan, and the Pacific, the first-to-file system applies, while in the United States the first-to-invent system applies. Patent applications in the United States must be filed within 1 year of the first offer for sale of the product or the patent filing will be void (Bastani and Fernandez, 2001).

Behfar Bastani and Dennis Fernandez, experts on intellectual property rights of nanotechnology (www.iploft.com), wrote on the multiple ways of defining protection of the invention idea. In addition to their study of patenting, the authors wrote on copyright as a way to protect the original expression of the idea, noting that copyright is less expensive and a quicker way to protect than patents. Copyrights offer legal protection immediately when the original subject idea is set in a physical medium that can be viewed. Copyright also encourages the owner to express the original idea in print or fixed medium. Copyrights are valid for the author's lifetime plus 50 years.

Their take on trademarks refers to the distinctive signature mark that can be used to protect the company, product, service, name, or symbol. While the trademark must not be descriptive or generic, Bastani and Fernandez (2001) note: "Legal protection is not offered to the technology, rather to the company goodwill and quality associated with the use of the recognized name or symbol." Compared to patents, trademarks can be obtained in a shorter time period, usually under 2 years, can provide exclusive rights within a region or nation, and as long as they are used commercially, may be renewed indefinitely. Trademarks frequently cost under $5,000 per registered mark.

Protecting your intellectual property is key—your creative genius is your moneymaker.

CASE STUDY: InvnTree

When a company can boast that 42% of its clients are U.S. and European patent attorneys, this is a strong indication that it meets the quality standards that are so critical in creating patents.

InvnTree is a people- and process-driven company that offers a wide array of patent services. InvnTree leverages its technology and patent expertise to provide superior quality patent services. InvnTree caters to a global clientele that includes law firms, in-house IP counsel, IP licensing and management companies, investors, start-ups, individual inventors, technology companies, and academic institutions.

Located in Bangalore, India, InvnTree leverages hand-picked talent from the vast pool of skilled professionals available in India. The team includes professionals pulled from prestigious institutes with advanced degrees in science and engineering who have strong experience in working with the Lean processes to deliver high-quality patenting services at competitive prices. Offering free patent strategy consulting to its clients, InvnTree uses a flexible service pricing model knowing that no two clients are the same, and that individual client needs will vary over time. Confidentiality is key. InvnTree incorporates a stringent legal, technical, and physical process to safeguard sensitive client information. The Bangalore site provides a low-cost location that significantly affects the price of patenting, allowing the technology company clients to take advantage of patent-related initiatives that they are often priced out of.

"Each of our team members either is an engineer or has an advanced degree in science. Additionally, they have thorough understanding of patent law. The combined skills enable us to exceed client expectation by delivering value added patent services. Further, we assign patent projects to the team that has the necessary qualification and experience in the technology field to which the patent projects relate. Hence, the insight provided by our team is highly valued by our clients. We follow an effective quality check process, such that the project execution is reviewed at various stages to ensure optimum quality."

Services

The excellence of the team at InvnTree has allowed for multiple patenting services, making it a one-stop center for patenting needs. Provided below is a list of patent services provided by the company:

1. Patent specification drafting
   - Complete patent specification drafting
     - The most critical document is the patent specification document. The patent specification document protects the interests in the invention and has to be drafted to perfection, requiring a thorough understanding of the technical aspects of the invention and a good understanding of the legal aspects of patenting. The experience and qualification in both technology and patent law allow InvnTree to draft first-rate value-added patent specifications.
     - The company has vast experience in drafting patent specifications, which are filed in patent offices that accept patent specifications drafted in English, such as the USPTO, EPO, and UK patent offices, and as PCT applications, among other patent offices.
   - Accelerated examination support document (AESD) preparation
     - This service is specific to clients who wish to file a patent application in the United States. Submission of AESD helps in expediting the patent examination process.

2. Patent filing and prosecution
   - Filing patent application
     - InvnTree has created an innovative business model that enables it to partner with patent agents/attorneys in many countries, as well as allow them to help clients file patent applications in other countries.
     - The model provides for the majority of the patenting activities to be carried out at the low-cost Bangalore, India, location, and only the filing is outsourced to the partner agents/attorneys.
     - The volume of business allows them to choose partners who offer filing services at competitive prices. By this

partnering the InvnTree business model results in significant cost savings to the client.
- The company also provides a service to help its clients file patent applications in USPTO and EPO themselves by providing the necessary support. This approach reduces patent filing costs significantly.
- Patent application prosecution support
  - Patent prosecution support to technology companies who have filed patent applications in major patent offices requires rigorous analysis and detailed drafting of responses to office actions/examination reports. This analysis and drafting is conducted at the Bangalore office and uses a business model similar to that followed in patent application, significantly reducing costs and ultimately benefiting the clients.

3. Prior art search and analysis
- Patentability search and analysis
  - The search and analysis service can be critical in helping clients make decisions with respect to filing a patent application for their invention, as it helps in ascertaining the probability of a patent grant for an invention; it helps to understand the scope that might be derived if a patent is granted; and it helps in drafting patent specification that may get granted with less hassle.
  - The search and analysis can identify features of the invention that may have the potential of a patent grant. Patent search strategies include an exhaustive search conducted in subscribed databases that cover patent data from more than 100 countries to identify patent documents that are relevant to the invention. A detailed analysis for each of the identified relevant documents is provided.
- Freedom to operate (FTO) search and analysis
  - A FTO search and analysis is an essential study conducted prior to introducing the product in the market and a proactive step toward justifying patent risks, such as patent infringement suits, by conducting a detailed patent search to identify relevant patents that are present in the jurisdiction of interest. A detailed analysis for each of the relevant patent documents is provided to the client.

- Patent validation/invalidation search and analysis
  - In the event of a possible patent litigation scenario or during patent litigation this search and analysis helps to invalidate/validate patents of concern by conducting a detailed patent search in subscribed databases that provide exhaustive patent data, identifying prior inventions that are relevant to the patent that has to be invalidated. By providing a detailed technical analysis on each prior reference, it allows the client a valuable tool in patent litigation
- Infringement search and analysis
  - The infringement search and analysis study conducts an exhaustive search to identify products/processes that might be infringing on the client's patent(s). A detailed technical analysis is provided to help clients take the appropriate actions to protect their patent rights.

4. Patent intelligence support
   - Patent landscape study
     - A patent landscape study is used as a knowledge base for technology trouble shooting, freedom to operate decisions, state-of-the-art updates, understanding competitor strategy, understanding technology trends, etc.
     - The study is an exhaustive search, analysis, and synthesis of patent documents relating to the technology domain of interest.
     - A detailed taxonomy of the technology is prepared in coordination with the client, based on the objective of the study. The patent documents are analyzed and arranged in the taxonomy to help the client obtain the most value from the available data.
   - Competitor portfolio study
     - A detailed study of a competitor's patent portfolio, the patent portfolio study is analyzed and broken down in a way that helps understand the competition better. A report is provided to the client with individual comment and analysis in reference to the client product, process, patent portfolio, or specific project requirement. The study is particularly useful in understanding competitors, mergers, and acquisition, determining patent filing trends, and identifying key innovators in a competitor's company, among other uses.

- The legal status of each of the patents in the portfolio can be provided.

5. Patent watch and alerts
    - Technology monitoring services
        - The technology monitoring service provides periodically updated updates of the most relevant information about the patent applications that get published and patents that are granted in the domain of interest. This is done by summarizing the results based on client requirement, so that it is easier and less time-consuming to make decisions.
    - Competitor monitoring
        - With the ability to monitor patent data in more than 100 jurisdictions, the competitor monitoring service provides a summarized report with periodic updates about a competitor's patent application publications and patent grants. This update report is invaluable to the client by providing information on filing pregrant and postgrant oppositions, monitoring developments made by competitors, predicting future products, and predicting a competitor's entry into newer markets.

Clients

Due to the professionalism, level of service affordability, and expediency, InvnTree has had tremendous success with outsourcing from around the world. The following is a list of client categories that InvnTree provides services to:

1. Academic institutions
    a. In helping academic institutions in building their patent portfolio, work is conducted at various stages of the patent life cycle. Due to the fact that patenting objectives of academic institutes are different from those of business/commercial establishments, services are customized to meet the requirements of academic institutes.
    b. Patent search and patentability assessment studies help pick inventions that can be taken forward for patenting. Providing patent specification drafting services helps the universities acquire patents for their inventions by assisting with application filing and prosecution services.
    c. An additional service provided helps to identify companies that might be interested in their patent portfolio.

2. In-house IP counsel
   a. In-house IP counsels outsource a variety of patent projects to InvnTree, typically for patent specification drafting, patentability searches, technology monitoring, competitor monitoring, patent landscaping, and patent prosecution support, among other patent services.
   b. By following proprietary Lean processes, quality checks, and incentive programs, it enables the delivery of high-quality reports at substantially lower prices within a short lead time, helping IP counsel make decisions based on rigorous research. With the outsourcing of patent research, expenses are substantially reduced.

3. Investors
   a. Patents are one of the key factors considered in valuing a technology company.
   b. Investors require insights to value a company's patent portfolio and are very interested in monitoring technology trends, which would enable them to identify investment opportunities.
   c. Customized research projects help investors' decision making.

4. IP licensing and management companies
   a. IP licensing and management companies typically require exhaustive patent due diligence to be conducted. These patent intelligence reports are generally time-consuming and labor-intensive. Because of the sensitivity and stakes involved, to carry out this comprehensive and exhaustive research, it must be done by subject matter experts. InvnTree uses subscribed databases that allow for patent research in more than 100 jurisdictions.
   b. The level of expertise needed and the time consumption of the project can be very expensive, but the company utilizes much of the patent searching capabilities of the Bangalore, India, location to help with cost.

5. Law firms
   a. InvnTree is one of the leading legal process outsourcing companies, which cater to a niche market of patent process outsourcing. Typically, patent attorneys initially use patent search outsourcing, and as the relationship grows, outsource allied patent services.

b. The most prominent advantage is the cost savings realized by legal process outsourcing, without compromising the quality of work. Law firms and patent attorneys reduce their cost structures through labor arbitrage.

c. The engagement model enables patent attorneys to scale up rapidly, providing access to the patent team that has expertise in several technical domains, thereby adding to the capabilities of the IP law firms. The company has a stringent quality check process to ensure high-quality work is delivered to the clients.

d. By outsourcing patent services that are time-consuming and laborious to the company in Bangalore, India, patent attorneys can concentrate on niche areas that interest them and are comparatively more profitable.

6. Start-ups and individual inventors

a. Understanding that individual inventors and start-ups work on shoestring budgets, InvnTree is interested in the start-up ecosystem and offers services at substantially subsidized prices, especially to individual inventors and early-stage start-ups, to keep them from compromising on quality by using the services of freelance patent consultants, some of whom have limited expertise. This start-up ecosystem mentality allows it to offer superior quality patenting services from a company that specializes in patent consulting.

b. The company frequently handholds start-ups in creating an effective IP strategy, which helps it in increasing their value, often providing free advice relating to various strategies that can be used to reduce expenses incurred while patenting an invention.

7. Technology companies

a. The most common patent services used by technology companies include patent search, patent intelligence, patent monitoring, and drafting services. The InvnTree team includes patent experts with diverse technology backgrounds. A patent project is allocated to a patent expert by matching his technology domain expertise with the technology domain to which the project relates. By using customized research that serves the objectives of technology companies, and following the proprietary Lean processes, quality checks, and incentive programs, high-quality reports at substantially lower prices within a short lead time can be delivered.

8. Engagement models
   a. On-demand model (ODM)
       i. By providing patent services on demand, clients can use the services when their needs arise. In this model, for each project, the scope and methodology are defined and a detailed project plan is sent to the client. Subsequently, the project is executed according to the plan. This model is suitable for clients who avail our patent services at discrete intervals.
   b. Dedicated hours model (DHM)
       i. In this model, the client can use the services for a predetermined number of hours every month. Based on the number of hours, the company offers services at a discounted hourly rate.
       ii. As opposed to the dedicated team model, in which a team has to be picked, in this model, the client has the flexibility of utilizing team members with diverse technical backgrounds. This model is suitable for clients for whom it is difficult to predetermine the technology domain of the work that would be outsourced.
   c. Dedicated team model (DTM)
       i. Providing a full-time equivalent team of patent experts dedicated to a client, the team size varies based on the requirement of the client, and typically ranges between 2 and 8 team members.
       ii. This engagement model is suitable for clients who wish to extend their bandwidth by maintaining an offshore team, which proves to be cost-effective.
       iii. In this model, the team, methodology, and other processes are customized to meet client-specific requirements. This model results in improved overall performance, quality, and value addition due to learning curve, while reducing costs.
       iv. This model enables clients to handle unanticipated workload, as there is a team dedicated to them.

Conclusion

Kartik Puttaiah is cofounder and CEO of InvnTree. The company was created with the vision "to be one of the most preferred patent service providers based on quality and cost-effectiveness."

From its ideas on start-up ecosystem to using the critical path method for its project management of client patenting, InvnTree can gain strategic advantage in the market for its clients by having the patent projects executed within the optimum duration and according to plan, allowing for cost efficiency. Its business models are strategically aligned with the customer needs and the utilization of the professional and technical education power that the area has; it has hit upon a model to reduce cost, provide jobs, help new business, and stimulate the economy in multiple arenas.

Using up to 90% of its workforce from the low-cost headquarters location of Bangalore, India, it has created an international company with an insignificant overhead, but high-level brain power of workers with advanced degrees from prestigious institutes. If that wasn't enough, this innovative company follows the Lean project execution processes, which additionally contribute to a reduced cost for services.

While many around the world are struggling to train the patent workers in new technology advances, InvnTree uses the vast technical power and patent expertise of the universities in the area that make technology a high priority.

By using a core staff of engineers and patent experts, this company can offer levels of comprehensive research that meets the needs of multiple entities. With subscribed databases, they cover patent data from 100 companies that allow them to execute patent research reports in a proprietary process they have created for analyzing data with the client focus to provide valuable insight that aids in well-informed decision making and legalities.

All of this leads to a shorter lead time in patenting that allows companies, individuals, universities, etc., the power they need to bring a patented product to commercialism.

(Information provided has been derived from a study conducted by InvnTree IP Services Pv. Ltd.)

Contact information for this company:

Kartik Puttaiah, CEO
Email address: kartik@InvnTree.com
Telephone numbers:
    U.S.: +1-408-5997304
    UK: +44-20-30265679
    India: P: +91-80-42124165; M: +91 98863 34262

Skype: Kartik.puttaiah
Website: www.InvnTree.com

5

Ethics

What about regard for ethics when it comes to inventions and cultural differences? Sociopolitical and sociocultural diversity have their own read on ethics. It is a touchy subject, but one that scientists have clearly taken to heart over the years when they defected and moved to other locations to bring their genius to life. New policies on Internet protocol (IP), trademark, and copyright offer hope, but still, cultural difference, sociopolitical factors, and of course, plain money-greed-power issues will play a part in this trillion dollar technology.

Our ethical values, for the most part, are ingrained and tightly tied to our societal norms, sociopolitical and socioeconomic conditions that often give an indication why some areas of the world are more corrupt than others. Transparency International (TI) (http://www.transparency.org/policy_research/surveys_indices/cpi/2010/results) is a global collation against corruption. The corruption perceptions index measures the perceived levels of public sector corruption in 178 countries around the world.

"With governments committing huge sums to tackle the most pressing problems, from the instability of financial markets to climate change and poverty, corruption remains an obstacle to achieving much needed progress." The TI Corruption Perception Index for 2010 shows that nearly ¾ of the 178 countries in the index score below 5: the scale runs from 10 (highly clean) to 0 (highly corrupt).

The top 10 rank: Denmark, New Zealand, Singapore, Finland, Sweden, Canada, Netherlands, Australia, Switzerland, and Norway. The 10 most corrupt are: Somalia, Myanmar, Afghanistan, Iraq, Uzbekistan, Turkmenistan, Sudan, Chad, Burundi, and Equatorial Guinea.

Those significant to this study of the commercialism of nanotechnology and our surrounding neighbors and threats are: Germany, 15; Japan, 17; United Kingdom, 20; United States, 22; Brazil, 69; China, 78; Mexico, 98; Nicaragua, 127; Iran, 146; Russia, 154; Venezuela, 164; Iraq, 175; and Afghanistan, 176.

Countries are looking to acquire new technology and become part of the nanotechnology wave for the vast economic impact it can have for them. The nanocompetition is on with states and countries striving for number one because of the vast amount of money and power that it will bring. As we become more global, our cultures come face-to-face in the workplace; diversities, ethics, cultures, and peoples will meld, similar to the way computers have opened up the lines between peoples, bringing new globally accepted ethics standards into place, but for whom?

K. ERIC DREXLER

Often described as the father of nano-technology, Eric Drexler set the technical direction for the field in his seminal 1981 paper in the *Proceedings of the National Academy of Sciences,* which established fundamental principles of molecular engineering and development paths to advanced nano-technologies. In his 1986 book, *Engines of Creation,* he introduced a broad audience to the fundamental technology objective: using machines that work at the molecular scale to restructure matter from the bottom up. Drexler's research in this field has been the basis for numerous journal articles and a comprehensive, physics-based analysis in his textbook *Nanosystems: Molecular Machinery, Manufacturing, and Computation.* In his publications and lectures, Dr. Drexler describes the implementation and applications of advanced nanotechnologies and shows how they can be used to solve, not merely delay, large-scale problems such as global warming.

Dr. Drexler served as chief technical advisor to Nanorex, a company developing design software for molecular engineering. In addition, he writes about nanotechnology and other topics on his blog, Metamodern. com. He has worked in collaboration with the World Wildlife Fund to explore advanced nanotechnology solutions to global issues such as energy and climate change. Recently, Drexler served as chief technical consultant to the Technology Roadmap for Productive Nanosystems, a project of the Battelle Memorial Institute and its participating U.S. national laboratories.

Drexler was awarded a PhD from the Massachusetts Institute of Technology in molecular nanotechnology (the first degree of its kind).

For more information on Dr. Drexler visit:

http://e-drexler.com/
http://metamodern.com/
Engines of Creation 2.0: The Coming Era of Nanotechnology—
    Updated and Expanded,
http://wowio.com/users/searchresults.asp?txtSearch=engines%
    20of%20creation%20by%20eric%20drexler

At this point we are looking to this wondrous new technology to coat our world, i.e., our roads, our clothes, our skin, our cars, etc., never once thinking of the miniscule particles and what havoc they may have on humans, our environment, and our ecosystem.

In our next chapter we will be covering the dangers associated to the human populace from using nanotechnology. Early nanopioneer K. Eric Drexler, who introduced the word *nanotechnology*, gave both the wonders and the dangers in a speech years ago. His dire prediction of things going wrong with the technology included a scenario of self-replicating nanorobots that malfunction and duplicate themselves trillions of times, rapidly consuming the entire world as they pull carbon from the environment to build more of themselves. It's called the *grey goo* scenario, where a synthetic nano-sized device replaces all organic material. Another scenario involves nanodevices made of organic material wiping out the earth and is called the "green goo."

Drexler's visionary work, *Engines of Creation*, originally published in 1986, is an important resource in the growing field of nanotechnology, detailing its potential applications and implications for society.

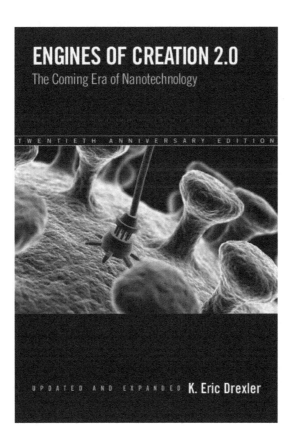

Coal and diamonds, sand and computer chips, cancer and healthy tissue: throughout history, variations in the arrangement of atoms have distinguished the cheap from the cherished, the diseased from the healthy. Arranged one way, atoms make up soil, air, and water; arranged another, they make up ripe strawberries. Arranged one way, they make up homes and fresh air; arranged another, they make up ash and smoke.

Our ability to arrange atoms lies at the foundation of technology. We have come far in our atom arranging, from chipping flint for arrowheads to machining aluminum for spaceships. We take pride in our technology, with our lifesaving drugs and desktop computers. Yet our spacecraft are still crude, our computers are still stupid, and the molecules in our tissues still slide into disorder, first destroying health, then life itself. For all our advances in arranging atoms, we still use primitive methods. With our present technology, we are still forced to handle atoms in unruly herds.

But the laws of nature leave plenty of room for progress, and the pressures of world competition are even now pushing us forward. For better or for worse, the greatest technological breakthrough in history is still to come.

Drexler's *Engines of Creation* prepares us for our future:

Advances in the technologies of medicine, space, computation, and production—and warfare—all depend on our ability to arrange atoms. With assemblers, we will be able to remake our world or destroy it. So at this point it seems wise to step back and look at the prospect as clearly as we can, so we can be sure that assemblers and nanotechnology are not a mere futurological mirage.

Two of the greatest threats/risks/fears that face the development of this technology are the possibilities of a catastrophic accident and misuse of the technology to harm. We only have to go back to the corruption index to know that the possibility and probability that this can and will happen as struggles for wealth and power exist.

In recent days of the earthquake and tsunami that devastated Japan, we go back to the dire warnings that went unheeded after the Chernobyl nuclear disaster and Three Mile Island. A global catastrophe is more than possible—it is probable with the technology we have if we do not listen to the warnings, create standards that protect all, and place caution over greed for money. As we venture further into the realm of nanotechnology, it is imperative and ethical that we pause to note the dangers and correct them before proceeding.

If we know there is a possibility, is it ethical to push forward to commercialize it without knowing what will happen? Is it ethical to put our plants in countries who need the work, not totally knowing of the health risks involved? Is it ethical for the wealthy to dictate the projection and direction of nanotechnology without the vote of the populace? Is it ethical to ignore the potential cures for some of our most devastating diseases based on fear of the unknown?

It brings the question back to this author: Why write a book to teach people how to move forward with commercialism if there is such an unpredictable future? And yet, that is precisely the answer. We rather need more people with an ethical bent to step forward and standardize and legitimize the production in ethical ways for the good of society and our planet, than to allow an open field for the ignorant of the technology and the unscrupulous who want to benefit financially at the cost of those less powerful or less informed.

Creating an open forum of how to navigate the technology and to progress ethically is the clarion call for all involved. The technology is here; it will do no good to hide your head in the sand and pretend it doesn't affect you. It will do no good to ignore it and wait like helpless puppies for the government to make your decisions on how it affects you. How we harness it and use it for the good of our planet and its people is society's calling.

In looking at the ethical issues facing society and how to create an action plan for the knowledge to be discriminated, it is imperative for our political leaders and those business owners in the nanocommunity to put ethics and society first. Each company would do well to create an *ethical decision making model* for commercialization similar to a SWOT analysis with professional, legal, ethical, and commercial issues addressed. The SWOTPLEC™ analysis creates policies to educate as well as to minimize the damaging effects of the technology on society. Knowing every aspect of your business gives you power to make solid decisions for the future.

Issues	Discussion
Strengths	Nanotechnology is at the stage where it will soon be able to build and commercialize complex molecular machines that can create virtually any product from the atom up. Satisfaction has been shown for the most part by society in the early developmental products, even though many did not know what they were using.
	It is an exciting technology that lets the mind expand with the potential for good in many industries.
	• Manufacturing: Material reuse of garbage broken to the molecular level and reassembled to new products, engines, etc. Precision manufacturing done at the microscopic level leaves little room for error, reducing waste, and bringing prices down.
	• Medicine: New drug creation and pathways for administering to cut down on side effects and toxicity. Nanomachine-assisted surgeries that can pinpoint to surgically remove the problems with microscopic repairs in hard-to-operate areas of the body to reduce excess trauma and speed recovery. Disease treatment, testing, etc., that will become less invasive.
	• Environment: Recycling and breaking down the garbage to the molecular levels. Toxic cleanup that will be more efficient, easier to do, create less hazard to the environment, and eliminate landfills that will reduce our natural resource consumption.

Issues	Discussion
Weaknesses	Many people at this point in 2011 have never heard of nanotechnology or know the benefits, challenges, and dangers of the technology. Nor do they realize they have been using nano-enhanced products in the form of their suntan lotions, cosmetics, clothing, automobiles, etc.
	Education of the masses. The Department of Commerce report done by the University of Illinois at Springfield spoke of the lack of readiness of our workforce and proposed that education in math and the sciences is desperately needed, as our workforce for nanotechnology will come from countries who are prepared and have more skilled labor forces.
	Patent system taking 2–3 years for turnaround, making some patents outdated before they are approved. This is a weakness for the funding, commercialization opportunities, and growth of the companies as they try to move forward.
	Development requires clean labs. Nanotechnology is a technology that requires special equipment and clean labs to develop their ideas and products. The clean labs and the equipment are extremely expensive, as an electron microscope may cost more than $100,000 itself.
	Venture capitalists need portfolios of "ripe" companies, ready to move forward quickly, to make themselves more competitive and able to operate effectively.
	Commercialism of the valley of death between universities and business to help new ideas be further developed rather than die in the lab.
Opportunities	Nanocenters in universities, government facilities, and R&D facilities give opportunity to small businesses that need access to the clean labs and technical equipment to develop and grow their products.
	Nanocenters are located around the world and offer varying opportunities as well as competitive opportunities for intellectual property.
	Governments around the world are pouring billions into nanotechnology.
	Nanotechnology is projected to be a trillion dollar business in just a few years, opening tremendous opportunities for economic rehabilitation and workforce needs.
	Nanotechnology knowledge and skills will see tremendous growth in the next few years.
	Nanotechnology can prove to be lucrative.
Threats	Unknown risks associated with the nanoparticles already in use and those that are being created.
	Risks involved and lack of standardization. A number of developments have gotten pushed through under the wire of standardization rules that do not actually represent the needs of nanotechnology. Also, there is a danger in the use of products from other countries that do not use the standardization of more ethical countries.
	Ignorance of the technology by society can cause misuse, distribution, working conditions, and disposal of potentially harmful substances and putting themselves in harm's way working with various products.
	Ignorance in how the technology can hurt.
	Corrupt countries, businesses, individuals, etc., that steal or ignore intellectual property rights because they are not honored internationally.
	Corruption that will use nanotechnology as a power for evil, such as biological warfare, eroding and violating the freedom and privacy of others, and enhanced drug delivery systems for illegal drugs.
	The grey goo and green goo scenarios of self-assembling nanobots and disassembling every molecule they encounter.

Issues	Discussion
Professional	Research currently funded by the government stakeholders via DARPA and NSF. Because the technology is being developed in so many different fields, how can all the needs be handled simultaneously via government? Other stakeholders are the researchers and their freedom of conducting the research they want. Users of specific nanotechnology products, and the general populace who may be affected because of usage and proximity. R&D being done in the colleges can secure patents and licensing for companies to help to get the technology commercialized.
Legal	The world will very shortly be in desperate need of laws regarding the litigations of nanotechnology. Nanolaw will need to become a reality to deal with the issues that arise, but who will create and enforce both local and international laws that will monitor the safe development of nanotechnology? What governing body will enforce them, both locally and internationally? To know the standardization rules and any laws that are on the books or coming on the books would help to create a nanotechnology business that could avoid accidents and litigation.
Ethical	Andrew Chen of Santa Clara University in the Silicon Valley described ethical issues: "Nanotechnology will give us more 'god-like' powers. It has the potential to eliminate other ethical issues (e.g., assembling beef instead of slaughtering cows, constructing cells rather than getting them from reproduction, etc.) … May lead to undetectable surveillance, Right to Privacy could be jeopardized. Do we have a duty to help and provide for others (countries) with this technology?"
Commercialism	Commercialism of the product will brush against ethics in how to protect our workforce, and consumers with safe products via packaging, and other information to inform. Nanotechnology is a perfect technology for the global economy and the sharing of key intellectual property to make the world a better place. Partnering and working with venture capitalists around the world will sometimes brush personal ethics when taking our production standards to corrupt countries. Will the success of the business or money make a business owner lower his or her standards and use the customs and standards of other countries to commercialize? Venture capitalists, government, and society can all be stakeholders in the commercialism effort and each require their due. Corporate pathology, when the rush of wealth surpasses the concern for the good of society, has been in the news the past few years as business practices sometimes become corrupt in the name of money. Preservation of the business goes back to ethics.

The last step is creating an action plan from the information included in the SWOTPLEC. The action plan would give criteria for continuing, regulating, and tagging technologies that can be tracked. The ethical decision action plan would also govern potential consequences of not going with the ethical plan, covering rights, fairness, and the common good for all involved. This ethical report would be an asset to the business plan, as well as help the company leaders to maintain control of their product and growth, circumventing potential problems that would cause them trouble.

Chen's ethics worksheet was an overall look at nanotechnology, not a specific company plan, and his results gave the following possible actions:

1. Nanotechnology R&D should be banned.
2. A nongovernment regulatory or advisory commission should be set up.
3. Adopt design guidelines:
   a. Nanomachines should only be specialized, not general purpose.
   b. Nanomachines should not be self-replicating.
   c. Nanomachines should not be made to use an abundant natural compound as fuel.
   d. Nanomachines should be tagged so that they are tracked.

Ethically, is it time for our populace to meet nanotechnology. As noted earlier in the text, many have been using products without knowing. Knowing the vast amount of good the technology brings without creating mass grey goo panic is a fine line in sharing the knowledge.

The Pennsylvania State University has an educational program called Nano4Me (www.nano4me.org) held at their NACK Center. Since 1998 the Penn State College of Engineering has been a national leader in nanotechnology education and workforce development and created the Nano4Me community of individuals that includes students, graduates, parents, educators, industry personnel, and government officials to serve the national educational workforce in the development of initiatives for nanotechnology. From 2001 until 2008, the Engineering College at University Park was home to the National Science Foundation (NSF) Regional Center for Nanofabrication Manufacturing Education. In 2008, through a grant from the NSF, the National Center for Nanotechnology Applications and Career Knowledge (NACK) Center was established.

The center is under the direction of Dr. Stephen Fonash, who holds the Bayard Dunkle Chair in Engineering Sciences and is director of the Penn State Center for Nanotechnology (CNEU), as well as director of the NSF Advanced Technical Education Center and the PA Nanofabrication Manufacturing Technology Partnership. Dr. Fonash holds 29 patents in his research areas, many of which are licensed to industry. He is world renowned for his work in solar cell physics, and his solar cell computer modeling code AMPS is used by nearly 800 groups worldwide. His leadership at Penn State has allowed the ethical development of nanotechnology to make the university a leading nanotechnology center.

The NACK Center is serious about its education of the populace and provides resources to educators teaching K–12 as well as professional development workshops and workshops of interest for the industry.

As Nanotechnology gains in strength and more people become aware of it, ethical groups will develop to voice the concerns. The Nanoethics Group (www.nanoethics.com) is a nonpartisan organization that studies the ethical and social implications of nanotechnology. Based at Cal Poly State University

San Luis Obispo, the group collaborates with nanotech ventures and research institutes on various ethical issues and innovations.

The Center for Responsible Nanotechnology (CRN) is a nonprofit think tank covering research and advocacy concerned with major social and environmental implications of advanced nanotechnology.

The International Council on Nanotechnology is an organization whose mission is to develop and communicate potential environmental and health risks of nanotechnology.

The Latin American Nanotechnology and Social Network (ReLANS) is working with academic institutions, government, and society to study, discuss, and inform on the political, economical, social, legal, ethical, and environmental implications of nanotechnologies that are developed locally as well as the collaboration with institutions and foreign centers and the imported goods containing nanocomponents.

Focus Nanotechnologi Africa, Inc. (FONAI) is a 501c3 nonprofit educational and scientific organization developed to combat "brain drain" and all forms of poverty, including science and technological poverty. It is active in the United States, Africa, and the Caribbean. Its mission is (1) education and training of the next generation of nanoexperts, (2) first-class physical capacity building, (3) research and discovery, and (4) industry and economic development.

Major accomplishments of FONA include:

1. Formation of the largest nanoscience and nanotechnology initiative: the U.S.-EU-Africa-Asia-Pacific and Caribbean Nanotechnology Initiative (USEACANI), with the first of a kind nanosciencetechacademy. Establishment of the *Journal Nanotechnology Progress International* (JONPI), the number one nanosciencetech journal in scope and coverage, with Nobel Prize winners as members of the editorial board. Online global nanosciencetech workshop held.

2. Networking for $800 billion EU power project in Africa for Europe, $50 billion each from the United States and EU for Africa climate change project, and G* $20 billion aid to Africa.

3. Membership to external advisory board of World Bank, $180 million project in Africa, and membership to steering committee EU-ICPC, $4.415 billion nanoproject until 2010, etc. (http://fonai.org/Home_Page.php).

USEACANI held a workshop/conference on April 24–28, 2011, at the LaGuardia Sheraton Hotel in New York City. Topics included nanoenergy (nanosolar cell, renewable energy), nanomedicine (HIV-AIDS, malaria, tumor, cancer), nanofiltration (nanofilm, nanofiber), nanosensor, nanoindustrial development, nanotech strategy, policy and funding, nanoeducation, nanoagriculture (production of disease-resistant plant or crop), nanoparticle, nanocomposite, etc. Guests: Nobel laureates, policy makers, presidents, private sector CEOs, etc., in the largest nanotechnology initiative.

CASE STUDY

The U.S. government recently reviewed the National Nanotechnology Initiative and used the guidelines below for review of the NNI (http://www.whitehouse.gov/sites/default/files/microsites/ostp/pcast-nano-report.pdf).

Statement of task for NNI and experts:

PRESIDENT'S COUNCIL OF ADVISORS ON SCIENCE AND TECHNOLOGY

Review of the National Nanotechnology Initiative

The 21st-Century Nanotechnology Research and Development Act of 2003 (Public Law 108-153) calls for a national nanotechnology advisory panel to periodically review the federal nanotechnology research and development program known as the National Nanotechnology Initiative. The President's Council of Advisors on Science and Technology is designated by executive order to serve as the NNAP.

The study intends to answer the following questions from the 21st-Century Nanotechnology Research and Development Act of 2003:

1. What are the trends and developments in nanotechnology science and engineering, as they relate to the NNI and generally?
2. Please describe NNI's progress in the last 2 years in implementing the NNI program.
3. Does the NNI program need to be revised? If so, how?
4. Is the balance correct among the components of the NNI program, including funding levels for the program component areas?
5. Have the component areas, priorities, and technical goals helped to maintain U.S. leadership in nanotechnology?
6. Have the management, coordination, implementation, and activities of the program been carried out appropriately over the last 2 years? How could improvements be made?
7. Have societal, ethical, legal, environmental, and workforce concerns been adequately addressed by the program during the last 2 years?

The study will address additional questions:

8. What should be the goals, priorities, and platforms for the NNI for the next 5 years in science, biomedical research, nanorenewables, and nanoelectronics?

Environment, health, and safety (EHS):

9. Has NNI established an acceptable strategy to address appropriate EHS priorities?
   - Is it effective?
   - Is it filling knowledge gaps?
   - Does it have an implementation and review plan?
   - Does it support evidence-informed decision making?
10. Is NNI appropriately invested in EHS research to address the priorities?
11. Has NNI implemented an approach toward achieving the goals of the strategic plan that leverages the strengths of federal agencies?
    - Have experts and other stakeholders outside the federal government been engaged effectively in addressing nano-EHS issues?
    - Have experts had input to the research strategy?
    - Have industry, consumers, and other stakeholders been engaged in identifying and addressing EHS issues?
    - Have state agencies been engaged in addressing EHS issues?
12. Has collaboration between federal agencies been effective at addressing nano-EHS issues?
    - Have cross-agency actions been coordinated or conflicting?
    - Have efforts to ensure interagency collaboration added value to the NNI?
    - Is there clear evidence of cross-agency collaboration having a significant impact on agency decisions and strategies?
13. Have federal agencies collaborated effectively with international partners on addressing nano-EHS issues?
    - Is there a clear and substantial awareness of international developments and actions?
    - Are federal agencies engaged with international partners, and are these engagements coordinated across the federal government?
    - Is there clear evidence for the United States influencing international developments, and benefiting from international collaborations?
14. Are there mechanisms in place to ensure research into the EHS implications of nanotechnology leads to evidence-informed decision making?
    - Is there a clear link between research and policy?

- Is research targeted to providing information regulators need to make informed decisions?
- Are there mechanisms to coordinate regulatory actions on nanotechnology across federal agencies?

Nanotechnology outputs:

15. What are the trends and developments in nanotechnology science and engineering and product introduction since 2008 (the last PCAST review): in the United States, internationally, and at the level of individual states?
16. Is the United States the leader, in what areas, and by what metrics?
17. Has progress been made since the last review in the transfer of science to products and processes, in technology transfer, and in the development of enabling standards?
18. Does an adequate science base to facilitate transfer of nanotechnology exist? If so, in what areas; if not, where are the gaps?
19. Has progress been made since the last review in the transfer of science to products and processes, in technology transfer, and in the development of enabling standards?

Program management:

20. Has appropriate progress been made since the last PCAST review in managing and implementing the program?
    - Has an appropriate strategic plan been developed for NNI broadly, and does it include societal, ethical, legal, and workforce objectives?
    - Do clear objectives exist to achieve the goals of the strategic plan?
    - Is the NNI appropriately balanced with regard to the components of the program, including funding levels for the program component areas?
    - What is the status of program management as it relates to coordination and implementation of efforts across NNI?
21. Is there a need to modify the strategic plan that was in place at the last review to reflect changing priorities or developments in the NNI?
22. Has the investment in equipment infrastructure been adequate, and has there been some larger coordination among agencies in developing that infrastructure?

23. Have the investments in education and communication to the public been effective?
    - What has been learned?
    - What more should be done?
24. If the societal, ethical, legal, and workforce objectives have not been well integrated into the program, how should NNI proceed in the future?
25. What is the status of program management as it relates to coordination and implementation of efforts across NNI?
    - Are there areas that require greater emphasis and coordination?
26. Does the NNI have the ability to quickly respond and adapt to new scientific/technological advances?
27. What is the impact of NNI beyond the creation of publications and patents?

Contributors to this report were the PCAST members from nine offices and agencies as follows:

1. Institute of Medicine
2. Office of Science and Technology Policy
3. Office of Management and Budget
4. Government Accountability Office
5. House of Representatives Committee on Science and Technology Staff
6. Senate Committee on Commerce, Science, and Transportation Staff
7. National Nanotechnology Coordination Office
8. National Research Council Staff
9. Woodrow Wilson International Center for Scholars

Discussion Topics

1. Do you feel that the nanotechnology initiatives need to voice and address ethical issues in more depth?
2. Do you feel they address the ethical concerns of the National Initiative?
3. What concerns were addressed, if any?
4. Where should these concerns be addressed?

6

Dangers

In presenting a case for the dangers of nanotechnology, we will do a dump of fears, theories, and projections that may or may not have the capacity to come to fruition. In most cases they will not because of the evolution of knowledge that keeps up with technology and the unlimited number of jobs that provide protection and security to the populace. That being said, the following are scattered bits of alarm in the minds of those viewing the dangerous side of nanotechnology.

Doctors are concerned that nanoparticles are so small that they could easily cross the blood-brain barrier—a membrane that protects the brain from harmful chemicals in the bloodstream.

Eric Drexler, the man who gave nanotechnology its name, taught us about "green goo" and "grey goo" of self-replicating machines. He also wrote about the human desire and how we manipulate technologies to meet our needs.

> When dog genes replicate, they often shuffle with those of other dogs that have been selected by people, who then select which puppies to keep and breed. Over the millennia, people have molded wolf-like beasts into greyhounds, toy poodles, dachshunds, and Saint Bernards. By selecting which genes survive, people have reshaped dogs in both body and temperament. Human desires have defined success for dog genes; other pressures have defined success for wolf genes.

Adding to the fear stirring in the minds, sci-fi authors and movie makers have brought the cryptic knowledge out of the protected catacombs of the universities. The public now watches scenes of terror, showing a cloud of dust that is in actuality an army of tiny computerized machines, or nanobots, controlled by the government. Michael Crichton's book *Prey* featured self-replicating nanobots that were accidentally released. The images stay in the minds, making many in the general populace around the world nervous about nanotechnology. Dean Koontz has a new nanotech bestseller out entitled *By the Light of the Moon*, with a new super race of nanobeings created from nanomachines that can devour humans and overthrow the "old race." *Lost Souls* is the fourth in his Frankenstein series.

High-tech surveillance, monitoring and tracking, military disassemblers, miniature weapons, and explosives. These meet the critera for causing a mass panic or a science fiction movie thriller. The fear is that if there were a problem with the disassemblers such as getting loose in the environment or self-replicating if there were a problem with their limiting mechanisms, they could have the power to multiply endlessly like viruses.

The fictional grey goo nanobots that are a result of molecular manufacturing gone wrong will bring about a real concern with molecular manufacturing in the next few years and the effects of crime and new ways to deal with it using crime science or early policing to detect crime using scientific ways to prevent it.

Terrorism, using new science technology, will create a different chaos than we are currently dealing with and will require new ways and new technology to deal with it. Kurzweil, in the article "Molecular Manufacturing and the Need for Crime Science," gave the following statistics: "In 2002, in the United States, police and detectives held 840,000 jobs and approximately 81% (680,400) were local-level law enforcers. According to the International Association of Chiefs of Police, based on facts from the Bureau of Justice Statistics, only 9% of local-level law enforcement agencies required their officers to have an associate's degree, and only 2% required them to have a bachelor's degree."

The trend toward uneducated law enforcement will need to change as our world changes. Education and knowledge will be the power and protection of the future to combat crime.

Nanotechnology brings with it strong social concerns for our safety and ultimate survival on this planet if the technology is misused or if there were massive accidents. We currently are studying ways to create more effective weapons for warfare, just as our enemies are. Medically nanotechnology has the power to re-create us, from rapid healing to making us stronger, smarter, and even enabling us with night vision. The question is: Are we still human or transhuman? How will this affect our evolution as a race and a society? Many think that this would create two kinds of races of people, i.e., wealthy modified humans and poor unaltered humans.

There are also fun thoughts, such as: Will there be a need for money in the future if everything we want to create could be done with a replicator? What do we do with the animals if we can produce a steak with a replicator? What about our economy, how will it be affected if molecular manufacturing becomes a reality? What happens to manufacturing jobs?

In 1993 Venor Vinge of the artificial intelligence community added his prediction to the already overworked minds fearing this new technology: "Within thirty years, we will have the technological means to create superhuman intelligence. Shortly after, the human era will be ended." Some fear it, others think we will all just upgrade to it, similar to computer or cellular upgrades.

One of the most demanding questions within the realm of nanotechnology is: What are the health risks and do individual governments have an adequate risk research strategy to ensure that it is being commercialized safely? We are just at the precipice with this new technology, and while nanotechnology was incorporated into more than $50 billion in manufactured goods by 2006, common sense rules that not all product and product enhancements were put through the rigors of a risk research program.

Kurzweil (2006) spoke to the rapid technological changes in his article "Nanotechnology Perceptions: A Review of Ultraprecision Engineering and

Nanotechnology." "The first half of the 21st century will be characterized by three overlapping revolutions—in Genetics, Nanotechnology, and Robotics (GNR). The deeply intertwined promise and peril of these technologies has led some serious thinkers to propose that we go very cautiously, possibly even to abandon them altogether."

A few years ago, computer maven Bill Joy (2000) wrote, "We are being propelled into a new century with no plan, no control, no brakes…. The only realistic alternative I see is relinquishment: to limit the development of the technologies that are too dangerous, by limiting our pursuit of certain kinds of knowledge."

"Technology has always been a mixed blessing, bringing us benefits such as longer and healthier lifespans, freedom from physical and mental drudgery, and many novel creative possibilities on the one hand, while introducing new dangers. Technology empowers both our creative and destructive natures."

Kurzweil notes that "most technology proponents have been arguing why relinquishment is impractical. They contend that the march of technology is relentless and we might as well go along for the ride, but with safeguards built in to make sure things don't get too crazy."

We are coming to a time of rapid acceleration in the field of nanotechnology while there is moderate governance. We are just a few years away from the projected $3 trillion in manufactured goods for 2014.

In the United States, Andrew D. Maynard, PhD, is one of the foremost international experts on addressing possible risks and developing safe nanotechnologies. Dr. Maynard is a chief science advisor for the Wilson Centers Project on Emerging Technologies. His recent congressional and public testimonies are available online at http://www.nanotechproject.org.

Looking at the cosmetic industry, for example, had Dr. Maynard checking the products currently on the market that have nanotechnology used in them. His walkabouts in stores gave him insight on many products. One such occasion took him to the beauty aisle that showed popular products using pro-tensium plus nanosomes of proretinol A. Nanosome technology utilizes the most practically effective tool for bidirectional delivery of both water- and oil-soluble ingredients in the skin. Other skin creams, sunscreens, etc., use particles of silica and nanoparticles of zinc oxide. One particular day cream, at $300/jar, uses buckyballs (buckyballs are carbon atoms a billionth of a meter wide) in its formula to prevent premature aging of the skin by acting as an antioxidant.

The concern is that when the particles get so small, they are likely to develop new chemical properties that may have unexpected risks. Research is lacking on the safety of nanoparticles and whether they can penetrate the skin; most scientists agree that the skin's tough outer layer is somewhat resistant to the nanoparticles, but research coming out of the National Institute of Environmental Health Sciences in North Carolina found that some nanoparticles can penetrate the skin and could potentially get into the bloodstream or interact with the immune system. This is an alert that more research needs done in this area, but is it mandated? When you have research done for ethical reasons and research done for potential commercialization, who chooses what to focus on? And if potential issues are not mandated research, how can society demand answers to its concerns? We are currently embroiled in a competitive nanotechnology R&D phase with venture capitalists wanting a rapid return on their investments.

There is currently a vacuum of regulation in the field of nanotechnology. One of the main reasons is that the money people or key stakeholders in the projects often do not believe it is necessary, while others think it is vital.

Once again it is the problem of educating. A large portion of the world's population has not heard of nanotechnology or its effects, good and bad. Having never heard of it, there is no way that they would understand that nanomaterials behave differently at the atom level as opposed to large-scale versions of the same substance. Our current standards and regulating boards were not created with the atomic level in mind, and so many existing regulations may not be applicable. Lack of regulation is a vital subject to liability insurers, and the insurance industry itself should lobby for clarity.

A clarion call has gone out for a more integrated risk-research framework that has more government control to manage environmental, and health and safety issues of many of the agencies. The EPA's responsibility and mandates are understanding the dangers with regard to health and environment from intentionally produced nanomaterials. We do not know at this point what the risks are of the life cycle of many of the products. We go back to the standards in the United States, which were not created for nanotechnology, but are the best that are currently used, and then look at other countries, whose nanoproducts we are already using in mass, and wonder if there has been due diligence toward risk before they were put on the market.

One concern is the similarity between asbestos fibers and carbon nanotubes. This similarity is being studied with great interest by the toxicology community because the carbon nanotubes have some of the same characteristics as the asbestos fibers with regard to shape, size, and biopersistency. Because of the lack of knowledge of the nanotubes at this point, there remains questions: Do they sloth off and can they penetrate the human skin? Are they released from clothing in the wash? How are the nanoparticles and nanotubes disposed of when the product has fulfilled its usefulness? The answer is, according to the University of Illinois Springfield report to the Department of Commerce, "the risk to the public, the environment and the workforce is presently unknown. It is recommended that factories and research laboratories treat manufactured nanoparticles and nanotubes as if they were hazardous and seek to reduce or remove them from waste streams."

Governments around the world need to work on the standardization to formulate toxicity testing protocols and screenings. At present there are not enough toxicologists to keep up with the speed at which various applications of nanotechnology are being developed, nor do many have the updated knowledge coming out of the lab to test the toxicity. Governments need to step up to the plate immediately to test and predict the impact of nanomaterials on humans, our food supplies, and the environment, as well as create educational programs for more toxicologists to be trained in nanotechnoogy. The urgent needs require that new laws and regulations protecting both the burgeoning technology and our people and environment need to be passed and enacted.

In looking at the risks and dangers involved with the production of nano-products, we review research from the prestigious Lloyd's—the world's leading insurance market, more commonly known as Lloyd's of London. Lloyd's has over 200 offices worldwide and insures more than movie star legs. As a leader in the insurance industry, Lloyd's will ultimately be the insurer of a portion of the companies and products and is researching the risks involved. The following information is from a Lloyd's emerging risks team report published in 2007.

LLOYD'S EMERGING RISKS

Nano particles—"The chemical reactivity of material is related to its surface area when compared to its volume. Dissecting a 1 centimeter cube of any material in to 1 nanometer cubes increases the total combined surface area some ten million times. Nano particles can therefore be much more reactive than larger volumes of the same substance." With this type of instability on such a small scale, the dangers of toxicity and mistakes that can be manufactured in large quantities are a concern.

Impacts on health—"It is unclear whether nanoparticles can cause chronic health effects." "Given that nano-sized objects tend to be more toxic than their large scale form it would be unwise to allow the unnecessary build up of nano particles within the body until the toxicological effects of that nanoparticle are known."

Unknown impacts on the environment—"Removing nanoparticles from the environment may also present a significant problem due to their small size. If absorbed, the particles may travel up the food chain to larger animals in a similar way to DDT though there is no evidence either way that this is a valid mechanism. There is still too little research in to the potential negative impacts of this technology on the environment." The study did state that there is evidence of some nanoparticles (copper or silver) that can be harmful to aquatic life. Ecotoxicology.

Positive effects of nanotechnology—"Because the benefits (of nanotechnology) are so seductive society may rush to capitalize on them before adequately assessing safety." Examples:

> Cars made to absorb more impact; earthquake, fire, flood-,
> and corrosion-resistant building materials; and even envi-
> ronmental cleanup made easier and cheaper with specialized
> nanoparticles.
>
> **Lack of regulation**—"Currently most all regulation of nanotech-
> nology is done using existing mechanisms. Stakeholders in
> nanotechnology are divided on whether specific regulation is
> required."
>
> **The Organization for Economic Cooperation and Development
> (OECD)** is an international organization that has released a
> "Nano Risk Framework" for risk managers. The EU is embrac-
> ing the approach, while the United States and Japan prefer a
> lighter regulation.
>
> www.lloyds.com/emergingrisks

Taking a closer look at the risks gives insight into the types of regulatory
processes that need to be developed. Lloyd's states that nanoparticles are free
particles that have a greater risk because:

- They are relatively cheap and can be manufactured in large quantities.
- They are already used in consumer products.
- Their properties can be very different than the larger forms of the
  material they are made from.
- They can be highly reactive.
- The particles often have unknown toxicity.
- Their toxicity can be difficult to quantify.
- They can disperse easily in air or water.

Industry needs to know at what levels of concentration are the nanoparticles hazardous to humans and what levels of concentration from multiple uses of nanoproducts could cause a serious buildup. Initial research
shows that some nanoparticles are acutely toxic when compared to larger
particles of the same material, such as ultra-fine carbon and diesel exhaust
particles. A second concern is the accumulation of nanoparticles over a long
period of time, and if there is a buildup from varying products. Much of
the danger associated with nanoparticles is still speculative, but because of
the multiple venues through which the nanoparticles can enter the body,
i.e., injection, absorption through the skin, inhalation, and ingestion, being
subjected to them is a concern since there is a correlation of infiltration
similarities of asbestos fibers and carbon nanotubes. "At the nanoscale,

particles can stick together (aggregate) or fuse (agglomerate) effectively creating larger particles."

An additional study concluded that nanoparticles "can penetrate intact skin at an occupationally relevant dose within the span of an average-length work day. These results suggest that skin is surprisingly permeable to nanomaterials."

Nanoparticles are currently being used in a vast number of products that have direct contact with the skin, i.e., cosmetics, sunscreen and clothing, etc. The question is whether, after absorption in the skin, they come into contact with the bloodstream and have reactions similar to those of toxicity collecting in certain cells of the body.

Lack of understanding the risks involved with nanotechnology has led to a regulatory gap with most countries. The EU and the United States are coordinating their efforts to regulate, while existing regulation in less developed Asian countries has not been specifically modified for nanotechnology.

In 2011 the commission will be required to respond to the European Parliament resolution that was adopted in 2009 on the regulatory aspects of nanomaterials.

Nevertheless, nanotechnology is here and will be developed. Looking through the commercialism lens, the risks open new opportunity for additional businesses and work in the future in areas such as nanolaw, cleanup, staffing, consulting, nanocenters, etc.

Molecular Self-Assembly

It's been called the holy grail of nanotechnology. It is sci-fi come to life, and making authors like Stephen King and Dean Koontz even richer: self-replicating nanobots. Fact or fiction? Fact. Scientists are looking at possible ways for bottom-up engineering to create the miniscule devices. Self-assembly involves one of the following: chemical assembly, physical assembly, or biological assembly. It is no longer a sci-fi idea, as researchers in the Empa team in Switzerland have grown a supramolecular chain. The chain is, of course, in a laboratory setting, but it has validated the idea (Figure 6.1).

Unscrupulous Individuals/Zealots

Earlier in this book we covered the corruption scale of countries. Because of the vast amount of money being poured into nanotechnology, this technology could inadvertently be the next seduction to the corrupt. Knowing that

FIGURE 6.1
Physical self-assembly. Carbon nanotubes assembled into a vertical orientation by an applied electric field. (From Y.C. Kim, K.H. Sohn, Y.M. Cho, and E.H. Yoo, 84(26), 5351.)

there is harm associated with it will direct illicit money makers to new ways of taking a piece of the action and passing on the liabilities and dangers to weaker, ignorant peoples looking for jobs.

Working on your brainchild, and hoping to make it thrive, can put savvy researchers at the mercy of the seduction of big money that promises to make your dreams come true—make the creative child of your genius come alive and help the world. Big money is talked about loudly to help you make a decision, while the fine print is spoken more softly. It is very frustrating as a researcher whose life has been huddled over a microscope in a lab now has to become instantly savvy in finance, marketing, IP, sales, and law to protect his or her efforts.

The Center for Responsible Nanotechnology is a non-profit think tank concerned with the major societal implications of advanced nanotechnology. CRN promotes public awareness and education, and the crafting of effective policy to maximize benefits and reduce dangers. "According to the reaction we get when we attend and speak at forums worldwide, interest in both opportunities and threats posed by exponential general-purpose molecular manufacturing is rising. We hope to contribute by engaging individuals and groups to focus on the real risks and benefits of the technology. Our goal is the creation of wise, comprehensive, and balanced plans for global management of this transformative technology." (Center for Responsible Nanotechnology)

CASE STUDY: NANO-ENHANCED ARMY

Daniel Moore, a developmental scientist and advisory board member of the Nanoethics Group, described the efforts of "the perfect soldier" in the article "The Nano-Enhanced Army." Moore described the nanotechnology factors as the latest potential tool for the military.

In 2002, the U.S. Army established an interdepartmental research center at the Massachusetts Institute of Technology (MIT), called the Institute for Soldier Nanotechnologies (ISN). The ISN is developing new ways to improve the survival and performance of U.S. soldiers via nanotechnology. The individual soldier is the

All technologies provide enhancement to humans, but that is not what is meant by *human enhancement*. This is still a grey area because of the lack of standardization and ethical rules in the nanotechnology realm. For this purpose we will use a partial list of the military technologies definition: purpose of use (defensive or offensive), level of use/impact (civilization or individual), type of use (shock or missile), amount of control required (human controlled or autonomous system), duration of use or impact (temporary or permanent), and the locus of use (internal or external).

It is vital to look at the problems that exist for the military with regards to human enhancement when designing military technologies. The following are a few that are being studied:

1. Soldiers need more effective tools in today's army to reach their goals.
   a. Nanoaluminum weaponry produces more powerful conventional explosives, faster moving missiles, and torpedoes that can bypass evasive actions.
2. Soldiers carry heavy loads, averaging just 100 lb in combat operations.
   a. Much of the weight is from equipment (communications equipment and power supplies—batteries).
   b. Nanoscale power generators utilizing piezoelectric nanowires that convert mechanical, vibrational, or hydraulic energy into electricity that can power the electrical systems carried by the soldiers.
3. Threats to soldiers come suddenly and unexpectedly and can cause injuries that cannot be healed quickly.
   a. Nanoscale materials made from flexible polymers and nanocomposites that can form nanoscale trusses—these can be woven into the battle suit, providing lightweight flexible body armor.

b. GE has an SiC matrix with nanoscale SiC fibers that can stop a bullet. Energy-absorbing and electromechanical materials act as an exoskeleton body that can apply casts, tourniquets, or provide CPR with no human input, as well as serving as exomuscles to augment the physical strength and movement of the soldier.

4. Sleep deprivation hurts a soldier's performance.
   a. Engineering humans who could fully function on 1.9 hours of sleep a night—what a giraffe needs.
   b. It is posited to lead to a twofold decrease in the casualty rate as the biggest factor in military effectiveness is "the degradation of performance under stressful conditions, particularly sleep deprivation."
   c. Sensors could also be used to detect fatigue in soldiers that could use metabolically dominant abilities that would allow soldiers to keep their cognitive abilities intact, not sleeping for weeks, all the while enduring constant extreme exertion by allowing targeted drug delivery systems to the individual site needed to provide chemical methods to enhance and control the mental states with minimal side effects.

Discussion Topics

1. Posttraumatic stress is one of the biggest problems of returning soldiers today. How do you think this will affect the emotional impact for the soldiers?

2. Do you think it is possible to bring an invincible superhuman back to live in the real everyday world of diapering babies, traffic, poodles, parades, and fireworks?

3. What do you see as possible concerns for the soldiers?

7

Standardization

Social capital is a widely used term referring to the *norms* and *networks* that enable a collective action. It is the societal cooperation of a people, enveloping relationships, institutions, and customs that shape a society's social interactions.

This societal cooperation is an "invisible hand" according to Adam Smith in his 1776 book *Wealth of Nations*. Smith posits that a man's pursuit of his individual interests is the criteria with which others pursue their own path. Rules of engagement must be established that motivate humans to take the concerns and needs of others into consideration when making decisions.

This invisible hand is the social capital. It is the connection between the social networks that govern society, both domestically and on the international level. As we become more global, there is an increasing need for the welfare of nations to make personal economic decisions, yet respect the needs and ambitions of other countries as they find their own economic balance. As we expand in the global market, international organizations are increasingly being called upon to work between the cultures, as a mediator of sorts, to create the greater community.

Never has a time to establish the invisible hand been more critical than now, when there is such an opportunity for man to create or destroy his world. The recent earthquake and subsequent tsunami and meltdown of the nuclear reactors in Japan bring it critically to forefront.

Each generation has shouldered the fears of destruction: Columbus falling off the end of the earth, or what he might find; Hitler's reign of terror; man's walk on the moon; Einstein's theory of relativity, and the subsequent atomic bomb; Elvis's gyrating new moves and rock music; the 1960s; the Internet; social media; terrorism; etc. Our forefathers have worried from the beginning of time about the dangers to society, but for the most part, the ideas had the luxury to evolve through multiple generations. But this time, technology is moving too fast to allow for the evolution of thought to filter through several generations to come up with a solution to the problem.

Moore's law states that technology will double every 18 months. Futurist Ray Kurzweil predicts that the next 20 years of technological progress will be equivalent to the entire 20th century—the equivalent of a century of knowledge in one generation by the year 2020. Today, we are citizens of the world, creating instantly what took generations to create before. We no longer have the luxury of time as a filter. Thus the invisible hand for nanotechnology needs to be *standardization*. We need to make sure that the decisions made are within the social context of human preservation and evolution.

Realizing the value of nanotechnology, the U.S. government, under the Clinton administration, raised nanoscale science and technology to the level of a federal initiative, calling it the National Nanotechnology Initiative (NNI). The recent presidential budget for 2011 shows a $1.8 billion investment, bringing the total cumulative investments in the past 10 years to nearly $14 billion.

This influx of government development cash and the subsequent economic potential is creating a "nanobuzz." Scientists, businesses, and venture capitalists understand that not all developments will make it to market, yet nanoenthusiasts are using the word *nano* as a convenient tag to attract funding. Pandora's box is open.

As with anything in life there is balance, yin and yang. What scientists are learning as well is that nanotechnology sometimes has unexpected consequences when manipulating properties at an atomic level. Issues such as drug resistance, rogue self-replicating nanomachines, pollution from the chemicals into the environment, etc., have created a critical awareness.

The wonderful power of nanotechnology to heal, as well as destroy, has led to an increased awareness on the societal level for how the world should deal with this new knowledge, i.e., social contracts on an international level.

Currently, the United States, Japan, Germany, and Russia are the four leading countries in nanotechnology research, with other emerging Asian countries of South Korea and China. Each world participant has developed much of its research independently, some in the form of corporate R&D, others under the watchful eye of government, and most through the intellect of the universities.

The U.S. focus, for example, is an integrated research system that has a close collaboration between universities, research institutes, and business. At this date, the United States is influencing the nano-related standards on the international scene through the American National Standards Institute's Nanotechnology Standards Panel (ANSI-NSP). This panel serves as the cross-sector coordinating body, facilitating the development of standards in the areas of testing, measurement, characterization, and terminology.

Japan, on the other hand, has led the nanotechnology standardization and risk effort through the Japan National Institute of Advanced Industrial Science and Technology (AIST). The AIST is an open forum that works with universities, government, national laboratories, media, industries, and businesses to address different aspects of nanotechnology on society, i.e., benefits, risks, standardization, and education. It holds monthly workshops in Tokyo.

Russia has its own ideas on standardization through the All-Russian Research Institute for Standardization and Certification in Mechanical Engineering. Formed in 2004, it is a federal executive body that focuses on technical regulating and metrology, whose responsibility is maintenance of instrumentation. In 2008, under the organization of the Russian Corporation of Nanotechnologies (RUSNANO) and the federal agency, Russia opened the first School of Metrology (science of weights and measures) and Standardization in Nanotechnologies and Nanoindustry. The school is creating a NANOCERTIFICA certification system for nanoindustry manufacture.

Germany and China have partnered in a bilateral forum: Frontier of Nanotechnology and Nanostandardization. The International Electrotechnical Commission (IEC), a leading global organization that prepares and publishes international standards for all electrical, electronic, and related technologies, reported in early 2010 that the Frontier of Nanotechnology and Nanostandardization has decided to focus on taking nanotechnology out of the lab and into the real commercial world.

These are just the top five countries and their focus on standardization, with varying ideas of what is most important. The positioning and power struggle will be a force with which to contend.

The exponential growth of nanotechnology makes it difficult to keep pace with the changes. David Rejeski is director of Foresight and Governance at the Woodrow Wilson International Center for Scholars in Washington, D.C. Rejeski feels that we have entered the next major industrial revolution, changing how we manufacture, where we manufacture, and even if we choose to manufacture. His approach to making any type of public policy in a Moore's law world would be with adaptation, coevolution, agility, and improvisation.

Rejeski, in an interview with *Ubiquity* magazine, stated, "We have entered the 21st century with outmoded bureaucratic structures firmly in place—structures designed to deal with the first industrial revolution and its aftermath, not proactively with the emerging knowledge economy."

This reflection on our own government says much about how we will define the social capital of nanotechnology as a nation. Taken on an international level, with countries just entering their first industrial revolution, this creates a complex arena for the standardization process. The power and money associated with this technology will bring about their own positioning among nations, as they accept, reject, and define the invisible hand of a global social capital.

The push is on. The nanotechnology die has been cast, there is no turning back; the social contracts are being drawn and the world power dynamics are shifting, as we take on this challenge as a global community. As citizens of the world, we ask ourselves: What do we owe each other with a technology standardization? Who is the deciding voice of this social contract?

Defining the invisible hand of social capital demands that we look at various international organizations and key players who currently set policy for the world.

The World Bank, as an example, uses five key dimensions as proxies for social capital: (1) groups and networks, (2) trust, (3) collective action, (4) social inclusion, and (5) information and communication. These key dimensions speak to the nanotechnology standardization criteria as well, establishing a foundation for negotiation.

At a recent nanoconference, the United Nations (UN) Subcommittee of Experts on the Transport of Dangerous Goods, and on the Globally Harmonized System of Classification and Labeling of Substances (GHS), spoke to the current activities of the International Organization for Standardization (ISO), the Organization for Economic Cooperation and Development (OECD), and the European Union program Registration, Evaluation, Authorization, and Restriction of Chemicals (REACH).

The UN subcommittee has proposed four categories of standards:

1. Terminology and nomenclature standards—To provide for a common language for scientific, technical, commercial, and regulatory processes.
2. Measurement and characterization standards—To provide an internationally accepted basis for quantitative scientific, commercial, and regulatory activities.
3. Health, safety, and environmental standards—To improve occupational safety, and consumer and environmental protection, promoting good practice in the production, use, and disposal of nanomaterials, nanotechnology products, and nanotechnology-enabled systems and products.
4. Specific applications of nanomaterials—Which lead to harmonized specifications.

As each international organization contributes to the implications of nanotechnology standardization, the organization most suited to be the leader is ASTM International, one of the largest voluntary standards development organizations in the world.

The ASTM, originally known as the American Society for Testing and Materials, was formed by scientists and engineers for the standardization of the railroad industry in the late 1800s. Over the past century it has remained a standards leader, switching to industrial, governmental, and environmental standardization requirements as the needs of society changed.

ASTM International has earned the trust for technical standards and is known for its high technical quality and market relevancy as it addresses the standards needs of the global market. The ASTM umbrella has even been used to solve the standardization challenges in such diverse industries as homeland security. In 2005, the ASTM created Committee E56 to monitor the growth in the nanotechnology arena.

Since most nanotechnology research is done independently in each country, creating a discourse among nations for the common ground of standardization often begins in the university, specifically with university resource spillover and the social capital network of the university scientist.

Dr. Ju Wang, of the Georgia Institute of Technology, posits that "the network size of a university scientist has a positive impact on the performance of a collaborating performance." He explains that the social capital of an individual university scientist is positively associated with regards to patents and venture capitalists because the network that a particular scientist carries with him or her provides channels which the respective partner is exposed and has access.

The scientists with more connections within their university, and those with vast national and international networking and publishing experience, provide more of a transference of knowledge, as opposed to those who have limited social capital. "Star scientists" are normally affiliated with major universities who can afford their salaries and research abilities. These scholars have a powerful influence in the technology community and can help to set the standards and will be the leaders of change to create the invisible hand of global standardization.

Adam Smith's invisible hand of social capital can lead to the "wealth of nations" in the new millennium. But we do not have the luxury of time as Smith did to evolve through the stages of development and to sit and ponder at the end of the day where this may lead. We are living the changes of our parent's entire century in just 20 years. Our Moore's law world demands our diligence and global networking to preevolve and create the standards of development for the future of the world (Figure 7.1).

Measurement is a major issue with regards to nanomaterials. To set future nanotechnology standards, potential regulatory issues will follow the successes of major successful nanotechnology players and standards committees. Some of the industry giants, such as IBM and Intel, are active on multiple committees that help them to define the future regulations for their nanotechnology projects.

FIGURE 7.1
Atomic force microscope (AFM) Veeco Modwel CP-II. Using high-speed electron microscopes enables researchers to look at materials on the molecular level.

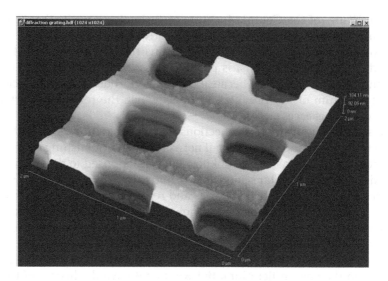

FIGURE 7.2
As the tip of the cantilever passes through the surface of the sample surface, it will create a two-dimensional image of the surface. In many cases, the difference in color gives a contrast about the topography of the surface. There is a legend of the image color palette next to the image (AFM picture) with information about the estimated depth based on the color intensities (dark vs. bright color). (From The Pennsylvania State University.)

Existing standards bodies are not specific to nanotechnology, but carry the best possible organizational structure to use when developing nano-specific standards. For example, the BSI Engineering Standards Committee began in 1901 and has grown into a leading global service organization that provides standard-based solutions to over 140 countries, developing private, national, and international standards; certifies management systems and products; provides testing and certification of products and services; provides training and information on standards and international trade; and provides performance management and supply chain management software solutions. The BSI Group is the national standards body of the United Kingdom and currently runs as an independent business that develops and sells standards and standardization solutions to meet the needs of the global business and society (Figures 7.2 to 7.4).

An additional standards group is the American Society for Testing and Materials (ASTM). ASTM was formed in 1898 by engineers and chemists from the Pennsylvania Railroad and became the ASTM International Society in 2001. ASTM International world headquarters is located in West Conshohocken, Pennsylvania. The society also has offices in Beijing, China, and Mexico City, Mexico. The ASTM has 141 technical standards writing committees. Today there are over 12,000 ASTM standards used around the world.

The following is a list of nanotechnology standards developed by ASTM (http://www.astm.org/Standards/nanotechnology-standards.html).

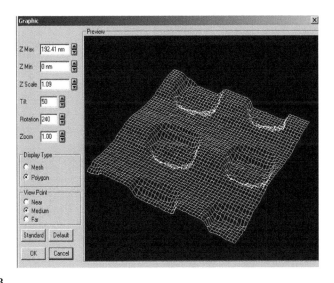

FIGURE 7.3
Now, the software can take this information and make a three-dimensional image. The first figure is a snapshot of the AFM computer software in order to set up a three-dimensional image (three-dimensional graphic setup .JPG). The next image is the three-dimensional picture that was obtained after all the parameters were specified in the AFM computer software (three-dimensional graphic .JPG). This second image is the one used on scientific papers and publications. The sample seems to be some type of a mesh array of nanoholes. You can make these micro- or nanoholes on a film by using wet-etching or dry-etching processes. Often, it is a good practice to know how much material was etched with a specific recipe as well as the quality of the etching (uniformity, surface roughness, angle of the side walls of the hole). (From The Pennsylvania State University .)

Characterization: Physical, Chemical, and Toxicological Properties

E2490-09	Standard Guide for Measurement of Particle Size Distribution of Nanomaterials in Suspension by Photon Correlation Spectroscopy (PCS) http://www.astm.org/Standards/E2490.htm
E2524-08	Standard Test Method for Analysis of Hemolytic Properties of Nanoparticles http://www.astm.org/Standards/E2524.htm
E2525-08	Standard Test Method for Evaluation of the Effect of Nanoparticulate Materials on the Formation of Mouse Granulocyte-Macrophage Colonies http://www.astm.org/Standards/E2525.htm
E2526-08	Standard Test Method for Evaluation of Cytotoxicity of Nanoparticulate Materials in Porcine Kidney Cells and Human Hepatocarcinoma Cells http://www.astm.org/Standards/E2526.htm

Environment, Health, and Safety

E2535-07	Standard Guide for Handling Unbound Engineered Nanoscale Particles in Occupational Settings http://www.astm.org/Standards/E2535.htm

Informatics and Terminology

E2456-06	Standard Terminology Relating to Nanotechnology http://www.astm.org/Standards/E2456.htm

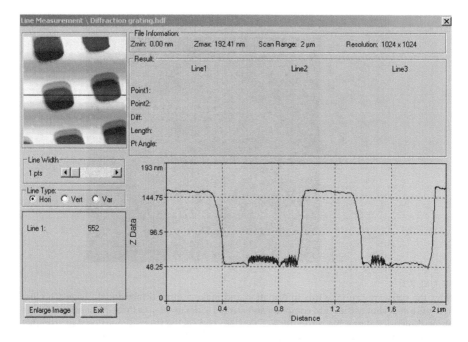

FIGURE 7.4
The last image is a snapshot of the AFM computer software screen in order to create a two-dimensional side view image (horizontal line measure .JPG). This image is obtained when the user slides the image (a reference line is place on top of the AFM image where the user would like to see a two-dimensional side view image). This is another way to see the depth and roughness of the sample features.

ASTM standards for nanotechnology provide guidance for nanotechnology and nanomaterials, as well as nanotechnology terminology, property testing, and issues of health and safety.

The American National Standards Institute (ANSI) established a nanotechnology standards panel in 2004 to enable those interested in nanotechnology standards to work together to coordinate the development of voluntary standards that include terminology and materials properties and measurement procedures to facilitate commercialization of applications and uses of nanotechnology. The American National Standards Institute–Nanotechnology Standards Panel (ANSI-NSP) membership is open to all parties interested in nanotechnology standards and provides a forum to aid those interested to define the needs, determine the work plans, and establish priorities for updating standards or creating new standards.

Taking the excellence of the current standardization organizations, future specific nanotechnology standards will follow three key approaches:

1. Specs of the production and the engineering practice of producing the end nanotechnology product
2. Defining how the finished nanoproduct can be integrated into existing supply chains
3. Defining how the technology will be integrated into the end product

CASE STUDY: RUSNANO, KAZYNA CAPITAL MANAGEMENT (REPUBLIC OF KAZAKHSTAN), VTB CAPITAL, AND I2BF

RUSNANO, Kazyna Capital Management (Republic of Kazakhstan), VTB Capital, and I2BF Holdings have recently signed a memorandum of intent to establish the Russian-Kazakhstan Nanotechnology Venture Fund. It is essentially the same idea behind the hedge fund. This fund's target size is $100 million. RUSNANO and Kazyna Capital Management, anchor investors for the fund, will each contribute $25 million. VTB Capital and I2BF Holdings will manage the fund's resources. I2BF Holdings is trying hard to attract private investments of $50 million to $100 million (NanoTech).

Expected to involve a broad range of economic sectors, the project will focus the funds on transferring cutting-edge technology, creating new forms of international collaboration, and stimulating development of financial infrastructure for nanotechnology markets. The venture fund, which is being established for 10 years, will draw investment resources to be used for projects in nanotechnology and the use of nanoindustry products in the Republic of Kazakhstan and the Russian Federation that meet the criteria for promising ventures.

"The attraction of two highly professional companies in the establishment of the venture fund is an excellent signal for other investors. Fusing different management approaches and technologies will help the management team attract additional investment to achieve targeted capitalization and ensure effective leadership for the fund," said RUSNANO managing director Dmitry Pimkin. "We have chosen management companies that can work effectively in selecting and supporting promising nanotechnology projects in Russia and Kazakhstan. Moreover, they will apply their experience in developing venture projects at the pre-industrial stage to maximize earnings for the fund's investors." Criteria for choosing projects for investment will be a forecasted return on investment, as well as a scientific and technical validity of the project.

Aidar Kaliev, head of Venture Investments at VTB Capital, claims that "VTB Capital is a leader in the Russian venture industry. Our experience in venture financing is considerable. There are currently five venture funds with aggregate value of 5.8 billion rubles under VTB Capital's direction. This year Russian Navigation Technologies, a VTB Capital portfolio company, made the first ever IPO offering in Russia. We believe that VTB Capital's global platform and experience will enable us to realize the potential of the Russian-Kazakh fund."

"Joint management of the venture fund is a unique approach in venture business. We are confident, however, that our cooperation will be successful. The I2BF team has worked for a long time in Western venture investment markets, monitoring their needs and developing the most farsighted company. We are prepared to bring the global expertise of I2BF to Russia, where we already see a host of technologies for which there is demand in world markets," adds Ilya Golubovich, managing partner at I2BF Holdings.

Kazyna Capital Management, registered on May 23, 2007, is a wholly owned subsidiary of Samruk-Kazyna, a joint-stock company of national (Kazakh) welfare. The company devotes its work to developing a market for direct investment in Kazakhstan. It has equity of 60 billion tenge (11.55 billion rubles); the total size of funds in which the company is an investor exceeds $2.8 billion. Kazyna Capital Management is a shareholder, along with industrial and other investors, in eight direct investment funds focusing on investment in Kazakhstan. KCM manages assets of about $1.5 billion (NanoTech).

VTB Capital offers the full spectrum of investment and banking products and services to Russian and foreign clients. It focuses its work on organizing issues of marketable debt and equity securities, developing direct investment business, conducting trading operations, managing assets and investments, and offering clients consulting services for mergers and acquisitions in Russia and abroad.

I2BF Holdings is an international asset management group engaged in venture financing and management of hedge funds in the United States, Europe, and Asia. I2BF was established in 2005. It is a diversified investment group specializing in high-technology sectors. The group has more than $100 million in assets under management. It directs venture funds exceeding $80 million. I2BF Holdings' hedge fund was a nominee for the 2009 EuroHedge Awards as the Best New Hedge Fund of the Year.

VTB Capital has attracted more than $40 billion into the Russian economy, in less than 2 years, organizing more than 90 transactions in the debt and equity capital markets. VTB Capital won the prestigious international award in the 2010 Euromoney for Best Russian Investment Bank in the Debt Market. The company's analytical department also garnered a number of international awards in 2010, including the Extel Pan-Europe 2010 and the All Russia Research Team (Institutional Investor) prizes (NanoTech).

8

Investors and Commercialization Centers

Studies show that one of the major problems of start-up companies is that the necessary and substantial cash is lacking early in the business venture. Monies needed for educated personnel, research and development, high processing costs, the long lead time for nanoproducts, and the lack of process scalability require a committed investment early on to make it to production.

The similarities of nanotechnology and the dot-com technology rise in the late 1900s are key in understanding the commercialization, as many dot-coms went bust because of lack of business skills and due diligence in the business setup. Mirroring the success of the dot-com technology market will benefit the nanotechnology focus on growth and funding. Governments are pouring funds into the technology, but potential investors need to know more, as many laypeople are not familiar with the complexities and future growth of the technology, as well as how to get involved. There needs to be a way to prove to potential investors that nanotechnology is a highly lucrative and marketable industry.

Start-up costs reflect cost of patenting, salaries for technical leaders, developers, salesmen, marketers, office workers, employee health costs, insurances, real estate, equipment, accountants, lawyers, executives, etc. The monthly costs can rapidly go into the hundreds of thousands of dollars. Getting an investor to buy into a new technology that will not reap payback for multiple years is a daunting task, even with the government stipends for nanotechnology development.

Many times a CEO's main focus is chasing the money. Working the field of investors, potential grants, and government funding is a full-time job to raise the money to keep the companies afloat. If the CEO is the scientist, needed for research and production, it is a twofold problem that often leads him or her to university and nanocenters for help.

Many avoidable problems with nanotechnology companies are their lack of business sense. As the nanoexperts spend their time and money trying to create a prototype, they flatline when it comes to the business aspects of their company, not knowing how to market their idea and network with potential investors. As their money is poured into their genius, they do not have money to hire an outside business manager (Figure 8.1).

Getting a venture capitalist to look at a struggling company that may take 2 to 3 years to achieve the patenting required in the early-stage development is not easy. Venture capitalists do not make their money waiting for great ideas to slowly ripen; they make their money in investments that will take 2 to 3 years to pay back.

Billions of Dollars

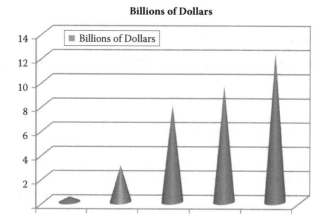

FIGURE 8.1

Nanotechnology investments from 1997 to 2006 from government, corporations, and venture capitalists.

The idea of using a hands-on angel investor who will assist in the business and marketing of the idea is one that is proving beneficial to all. Angel investors normally opt to invest, but also advise the company in business basics as well as introduce the idea to the correct people and make it potentially investable.

Utilizing this strategy is one way to bridge the gap as the technology company would achieve more success in the early stages of development. If more nanotechnology companies can transcend the initial economic difficulties, the success ratio could attract venture capitalists and other high-volume investors to take on a longer investment with nanotechnology companies than they are typically known for making in other industries. The key for success is lowering the percentage of failures in early-stage development, cutting the risk for interested investors. Initial angels who originally invested during this critical time could be financially rewarded for providing one means to bridge the gap.

Start-up nanotechnology companies frequently need what is called gap funding to help get through those first critical years waiting for the patents and R&D. Gap funding is leveraging federal and state funds with private funding. Business start-ups often realize the large federal funding in the form of perks set up at universities who received the grants that teach business entrepreneurial skills or through research and patent protection and the support and use of clean rooms, labs, equipment, and nanotools.

The Small Business Innovation Research (SBIR) funding model is a set-aside program for domestic small businesses with potential for commercialization to engage in R&D. Many nano start-ups are interested in securing gap funding from the SBIR program and become disillusioned to find that the model does not fit the time requirements of today's nanotechnology businesses, which take sometimes 2 to 3 years for R&D and another 5 to 7 years to complete the production.

Finding angel investors has been the recent trend in funding of nanotechnology. As mentioned earlier, angel investors typically invest money during the early stages of development, teaching scientists and technology experts smart business sense. It is in this way that angel investors help themselves by mentoring the very company they invested in, often involving themselves in each aspect of the company. By becoming so involved in the interworking of the companies in which they invest, success of the company is of primary importance, resulting in the investors' willing to stick with the company through the entire development process. Unlike angel investors, venture capitalists are not involved in the business aspect of the company. Their only concern is the business's success. Angel investors have more than just money invested and, for that reason, want to see a company succeed for more than just the monetary return on their original investment. Because of this hands-on structure, angel investors will be more inclined to go out and seek more funding if it is needed later in the nanotechnology process.

Recently, angel investors are becoming extremely harder to please, making it difficult to receive funding. Both the downturn of the economy and the dot-com crisis have left these investors weary. Even though the angel investors have picked up a large amount of slack left by the now wary venture capitalist industry, they are hesitant to invest unless they are convinced the idea is a very lucrative one. Angel investors have invested billions of dollars into the early-stage company sector. Venture capitalists have not even come close. Angel investors were known for taking risks based on instinct and guts; more recently, they rely on the facts. It is not a surprise that the shift from guts to facts dates back precisely to the dot-com crash. When the dot-com crash happened, all investors became much more risk-averse. This resulted in angel investors wanting to see a positive cash flow within the first 18 months of making their investments. This type of quick return is unlikely unless an idea for a marketable product is already implemented (Dahl).

In one way, receiving investments from angel investors has become much easier. Until recently, there was no clearinghouse for angel investors, which made it very difficult to locate them. The Ewing Marion Kauffman Foundation created the first angel investor network called the Angel Capital Association. Its goal is to document a large number of angel groups sorted by state and region (Dahl).

Investing in nanotechnology is much different than investing in most other industries because this type of company is relatively new and innovative; there is limited criteria upon which to make an investment. Investors are weighing the advantage of early-stage nanotechnology development. Angel investors can be a quick relief for the initial funding of a company, but eventually that company will need to secure additional funding after initial development. The appeal of angel investors to nanotechnology companies is that they are willing to invest in small companies with smaller markets.

Angel investors are much different than venture capitalists. Most venture capitalists invest other people's money and make their profit by taking a percentage of the return. A venture capitalist fund manager does not risk any of his or her money. This makes it easier to take risks and in some cases make poor decisions. Conversely, angel investors are investing their own money. This makes them more invested in the success of each investment and each company. This also allows them the freedom in choosing which market in which to invest. Additionally, they do not need as high of a return on their investment as venture capitalists, making it feasible for them to invest in a small company or a small market. Angel investors are very involved and carefully choose which companies have the best management teams. Due to the fact that angel investors have limited funds that they are willing to invest, they need to know exactly when a company will be able to make money or raise additional funding, if necessary.

According to Abrams, there are five main things that angels need to have in order to invest:

1. The company they are investing in must have an idea that they understand.
2. Ideally, it must be a business from an industry with which they have been associated. It must be something they can become passionate about.
3. Angel investors do not need to invest. They have enough money to be a savvy investor, already making it probable that they are well off enough financially to not invest in each idea.
4. They must trust the management. They must respect the management and like the management.
5. The main factor is that angel investors need to be able to bring added value to the company. They do not want to invest huge sums of money or invest more money later.

Economic hardship the past few years has added to the tightening and scrutiny of investing. The confusion over relative roles that venture capitalists and angel investors play is understandable because, lately, there has been a grey area in investing that has not been present in a long time. Angel investors are reviewing potential deals very carefully. They are studying business plans, requiring mature management teams, and doing their due diligence. The business plan proposed to the angel investors needs to be written to meet their needs and wants perfectly. It is beneficial to research potential investors prior to building a personalized business plan for each of them. Because so many more people are soliciting angel investors than ever before, it is imperative that the company appeals to the angels because there is so much competition trying to gain funding for start-up costs, research and development, and multiple nanotechnology patents. Recently, angel investors are more willing to expand their timeline from 18 months to 5–7 years. This is great news for the nanotechnology industry (Brown).

Bridging the gap from the research lab to the commercialization of the product is requiring that nanotechnology companies look toward new financial and investment trends. Even with innovative and life-saving discoveries, there is a high percentage of nanotechnology companies that never make it. Investors are waiting for other investors to take the first step in early-stage development. They want to wait and see if it is profitable and then jump in while it's still very lucrative. It is likely that one day, many investors will look back and regret the fact that they did not jump in on the nanotechnology opportunity sooner.

Venture capitalists have been wary of taking large risks ever since the dot-com crisis of the early 2000s. Venture capitalists are those few with large sums of money that are willing to invest, mainly driven in their investments by timing. These investors wait until the target company is in need of their money. By timing these investments, they are able to maximize their return in the shortest amount of time. They are generally not interested in funding companies during the idea stage of development. Venture capitalists invest once the company with the idea removes some of the risk by developing prototypes and buying patents, ensuring that the company already has beta customers and is gaining revenue from the idea.

This type of preinvestment development is very uncommon in nanotechnology, which is why venture capitalists do not usually make big investments backing nanotechnology ideas, but could then be sought to make an investment once the company is up and running and the initial risk has been substantially lowered. Venture capitalists swing for the fences. They need to see large returns because only a small percentage of their investments pan out and become fruitful. Venture capitalists generally invest more than $1 million into a company; however, they fail on about 80% of their investments. They break even on about 10%. In order to turn a profit, they need the remaining 10% of their investments to show a 10 times or 1,000% return on investment. Venture capitalists usually want to see a return in 5 to 7 years. Because of the time it takes for nanotechnology to go from idea to a fully functioning fruitful company, this adds yet another layer, making it difficult for venture capitalists and nanotechnologists to cooperate (Lavinsky).

The nanotechnology company can maximize this type of investment by making multiple investors believe in their particular company, creating an environment of trust that can build into a solid relationship before receiving a VC investment. Because of the multiple different companies going to venture capitalists every day looking for funding, the best way to attract a VC is to show him or her that other competing investors are also interested. When an investor becomes aware that there is other interest in a particular company, it gives that company the leverage of competition. This creates a scenario where competing VCs become willing to increase the amount of their investment or be more eager to invest (Lavinsky).

The venture capital market is going global. Venture funds collectively increased their investments by 19% over 2009 to $21.8 billion. The number of deals grew by 12% to 3,277. Even though growth slowed in the last two quarters of last year, it was still the venture capitalist industry's first positive year-to-year growth since 2007. There was a 30% increase in funding of early-stage companies. Software companies were given the largest piece of the venture pie last year, with $4 billion spread across 835 firms (Delevett).

It can be projected that the investments made by the venture capitalists are becoming more technology related, and this bodes well for nanotechnology companies. The future is bright, and once these companies surpass the valley of death stage, the return on investment will convince other investors to consider early stage.

The following chart shows the amount of money spent by venture capitalists in the last four years. After the dot-com crisis, venture capitalists were very stingy with their money. In 2005 and 2006, there was an increase in venture capitalist investing. Researchers assumed that in the following years this increase would continue, but this assumption proved to be false with the volatile economy and subsequent crash in 2008. Finance and market projectionists have reason to believe that it will be a few years before the venture capitalist market turns back around (Figure 8.2).

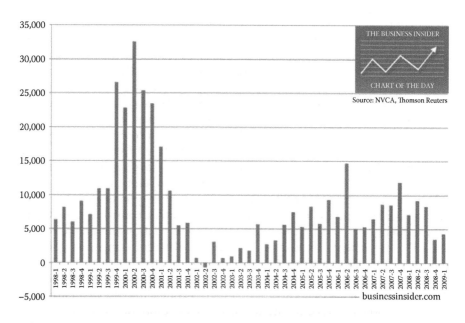

FIGURE 8.2
Venture Funds Raised 14 percent less in 2010, falling to a seven-year low.

Forbes' online website recently posted its view for the noticeable contraction of the venture capital industry:

> The numbers clearly point to a downward sloping line for venture fundraising, both quarter by quarter and year over year. Thirty-five funds raised nearly $3 billion in fourth quarter of 2010, which is down 6% from the third quarter 2010, when 49 funds raised $3.2 billion. Overall in 2010, 157 venture capital funds raised $12.3 billion, which the release points out is the fourth consecutive year of declines and the slowest annual period for venture capital fundraising since 2003. (Farrell, http://blogs.forbe s.com/maureenfarrell/ 2011/01/18/venture-c apital-fundraising- continues-its-downward-slide/)

NVCA/Thomas Reuters numbers can be viewed online at NVCAccess. NVCAccess is heralded by the company as the official voice of venture capital and keeps a close eye on the research and trends. A recent update on May 18 had the fourth quarter of 2010 showing some encouragement as venture returns improved at the end of the year, marking the second consecutive quarter that showed marked increases (http://nvcaccess.nvca.org/index. php/topics/research-and-trends.html).

Although venture capitalist investments have increased in the world market, in the United States they have been declining. There is no reason to believe that this trend will change in the near future.

With each passing year the nanotechnology market is becoming more commercialized. A proactive approach to funding this type of market will help this commercialization process happen much more quickly.

Many large companies, wanting to get away from their own costly R&D, are very interested in absorbing a proven product in the form of a small business start-up. Working and funding start-ups with seed money gives large companies a chance to focus on niche markets, allowing the start-up to focus on concentrated research and refine the product. This liaison is of very little risk for the corporation, but potential risk for the start-up, which can be absorbed into the larger corporation, which can take its IP and market. If the start-up fails, the corporation is not liable for any legal battles.

Another avenue for start-ups to obtain the funding needed is the services offered at university and government nanocenters. The Project on Emerging Nanotechnologies (PEN) reported that in the United States alone, all 50 states, including the District of Columbia, are represented by 1,200 university, government laboratory, company, or nanotechnology organizations involved in nanotechnology R&D and commercialism. This number increased significantly—over 50% in 2 years. Additional findings identified over 182 universities and laboratories working in the field of nanotechnology, with the top three sectors as medicine and health, tools and instruments, and materials.

Helping to breach the gap, (valley of death) between laboratories and commercialism is becoming an industry in and of itself as government-funded centers are springing up in multiple states and countries. These centers work with ideas that have come of age and are ready to take through the stages of commercialism, offering start-up grants and funding for chosen businesses.

One such business is the Pennsylvania NanoMaterials Commercialization Center, located in Pittsburgh, Pennsylvania. The center offers nanomaterials researchers an opportunity to partner with companies to commercialize the research. It offers companies the opportunity to forego or accelerate their product development by partnering with researchers and it allows investors to explore opportunities with existing nanocompanies.

This year the commercialization center is making available $980,000 in grant funds to Pennsylvania university researchers; both small and large companies to support advanced nano R&D; and manufacturing in the commercial and defense sectors. Executive Director Dr. Alan Brown admits that the collaborative approach seems to make the difference in accelerating the development of the innovative technologies.

The center uses the funding by providing $30,000 grants for fast-track commercialization projects and $200,000 grants for full commercialization projects.

The Pennsylvania nanocenter began in 2006 under the direction of the Pittsburgh Technology Council via a consortium of four western Pennsylvania companies: Alcoa Technology, Bayer MaterialScience, PPG Industries, and U.S. Steel. It has expanded to partner with Carnegie Mellon University, University of Pittsburgh, The Pennsylvania State University, Lehigh University, the Department of Community and Economic Development for the Commonwealth of Pennsylvania, and Air Force Research Labs, as well as over 300 companies, organizations, and individuals that are involved in some aspect of nanotechnology.

Smart System Technology and Commercialization Center (STC)

The Smart System Technology and Commercialization Center (STC) was created in 2010 via a merger of two of New York State's Centers of Excellence: Infotonics Technology Center (ITC) in Canandaigua and the Center of Excellence in Nanoelectronics and Nanotechnology at the CNSE.

The STC facility is located outside of Rochester, New York, in a 140,000-square-foot facility that features more than 50,000 square feet of certified clean room space with 150-millimeter wafer production, plus an 8,000-square-foot MEMS and optoelectronic packaging facility.

Empire State Development (ESD) and NYSTAR will invest up to $10 million in STC, which will be managed and supported by CNSE, positioning New York State as a global leader in smart system and smart device innovation and manufacturing.

World Premier International (WPI) Research Center Initiative

The World Premier International (WPI) Research Center Initiative was launched by Japan's Ministry of Education, Culture, Sports, Science, and Technology (MEXT) in 2007, when NIMS and four national universities, University of Tokyo, Kyoto University, Tohoku University, and Osaka University, were selected for grants. MANA, the international center for materials nanoarchitectonics, was created that same year. The centers have a concentrated support for projects and relatively high standards, giving them a highly visible presence in the global scientific community that allows them to attract frontline researchers around the world to come and work in the centers. Currently the WPI has six centers, with Kyushu University added in 2010.

University Center	WPI Research Center	Research Field
Tohoku University	Advanced Institute for Materials Research (AIMR) www.wpi-aimr.tohoku.ac.jp	Material science
University of Tokyo	Institute for Physics and Mathematics of the Universe (IPMU) www.ipmu.jp	Astrophysics
Kyoto University	Institute for Integrated Cell-Material Sciences (iCeMS) http://www.icems.kyoto-u.ac.jp/e/	Meso-control and stem cells
Osaka University	Immunology Frontier Research Center (IFReC) http://www.ifrec.osaka-u.ac.jp/index-e.php	Immunology
National Institute for Material Sciences	International Center for Materials Nanoarchitectonics (MANA) (NIMS) http://www.nims.go.jp/eng/	Nanotechnology and materials science
Kyushu University	Carbon-Neutral Energy Research Institute http://www.kyushu-u.ac.jp/english/index.php	Energy and environment

Japanese National Institute for Materials Science (NIMS) International Center for Materials Nanoarchitectonics (MANA)

MANA is one of the five research centers that make up the World Premier International (WPI) Research Center Initiative. MANA's goal is to become a "melting pot" drawing top-level researchers from around the world to promote interdisciplinary research via materials nanoarchitectonics. The center is targeting brilliant and innovative young scientists. Part of the MANA melting pot mission is to construct a network of nanotechnology centers around the world.

One such center is the University College of London (UCL), which, in a collaborative agreement with the Japanese government, created a major new nanoscience research center. The Cambridge University Nanoscience Center is just one of the four satellites of MANA that are based outside of Japan; the others are in America and France.

The UCL/Cambridge University partnership is slated to become a world-class research center, expected to pull together the expertise of domestic scientists and their counterparts from the Japanese National Institute for Material Science (NIMS).

UCL Center for Nanotechnology, UK (www.lcn.ucl.ac.uk/)

The UCL Nanotechnology Center is proposed to act as a focal point for interdisciplinary nanoscale materials and device research at UCL, bridging biomedical, physical, and electrical sciences with a focus on the chip-to-cell interface, thought to be an essential step to remain internationally competitive in biotechnology.

The UCL Nanotechnology Center draws experts in the Departments of Electronic and Electrical Engineering, Physics and Astronomy, and Medicine as coapplicants, bringing together in a single location scientists with expertise in theory, hard and soft condensed matter physics, and electrical engineering.

Part 2 of this book provides lists of resources on various subjects. Additional university and government centers are available for research into a center close to your business.

The Pennsylvania State University

The Penn State Research Organization is the licensing arm of The Pennsylvania State University, and by 2005 had helped to license more than 50 nanotechnology discoveries, which have included nanoparticles for drug delivery and thin-film nanostructures for medical diagnosis. In the 10 years since its inception over 1,000 companies, including industrial giants DuPont, Dow, Johnson & Johnson, and PP&G, have sponsored Penn State researchers. Penn State has helped to launch 11 nanotechnology companies.

As one of the original land-grant universities, Penn State's charge is "to seek the full and rapid dissemination of the creative and scholarly works of its faculty, staff, and students in order to provide timely benefits to the citizens of the Commonwealth and the Nation."

Penn State partners with and utilizes the services of the following business development offices to aid innovators toward commerciality with their ideas.

- Innovation Park: A 118-acre research park near the main campus. Provides space, access to Penn State researchers and facilities, and business support services that help companies transfer the knowledge within the university to the marketplace.
- Industrial Research Office (IRO): Works to promote university-industry research partnerships, and assist companies in identifying and accessing Penn State faculty expertise and research centers.
- Penn State Intellectual Property Office: This office manages the discoveries and inventions with commercial potential of university researchers.
- Research Development Office: The research development office assists Penn State faculty and staff with the start-up of companies based on Penn State research.
- Ben Franklin Technology: Partners with Penn State to provide small to mid-size high-technology businesses with funding and university expertise.
- PENNTAP: Supports technology-based economic development by helping Pennsylvania companies improve competitiveness by providing a limited amount of free technology assistance to help resolve specific technical needs.

CASE STUDY: DISASTER RELIEF INFORMATION FROM TOHOKU UNIVERSITY

CASE STUDY – HEDGE FUND OPTIONS FOR NANOTECHNOLOGY
Mark Charleston Case Study

Mark Charleston was a senior at the The Pennsylvania State University in 2010. He was required by the business course of study at the campus he attended to complete an internship or research project to graduate. Mark chose 495 C, a research-style internship. He is very interested in finance and wanted to learn more about hedge funding. His first inclination was to do it on a hedge fund business for himself and his brother for commercial purposes. With any internship, it is the duty of the student to find a business to supervise his or her internship. When Mark approached Dr. Sparks, they spoke about the hedge fund for personal purposes, but after reflection, and her discussion of this book on nanotechnology, his new focus for research centered on the entry of businesses striving for commercialism and the money issues to cover the initial costs of setting up the business before they can attract financial experts such as angel investors or venture capitalists. Mark took a great interest in hedge funds and delved into the mysteries of nanotechnology, opening a new commercial outlook for himself in the future. Mark graduated in May 2010 and has taken a position as a mortgage broker while he continues his research into hedge funding and nanotechnology.

The following is an excerpt of his research:

Nanotechnology is growing exponentially every day. However, there is a severe lack of funding for this growing enterprise. In a market where investors are unsure of a definitive timeline for earning a return on their investments, they are naturally wary of tying up their funds in a project that has no guarantees. Despite their apprehension to invest, savvy investors understand the tremendous potential of nanotechnology. Many believe it will soon take over the technology world. Nanotechnology can improve everything from consumer products to cancer curing prototypes. Unfortunately, these ideas are falling into the Valley of Death. This is the obstacle that the nanotechnology field is working hard to overcome.

Many scientists are working to create new ideas in the nanotechnology field, but progress is halted when funds are unavailable to propel them into reality. Venture capitalists will not invest because of the uncertain timeline for returns on potential investments. Angel investors can be a temporary fix, but inevitably lead to wasted time and money since their investments prove to be smaller than the amount needed to bridge the valley of death. There will still be a need for venture capitalists because the majority of angel investors do not have infinite funds to invest. The objective is to create a way where investors' risk is lowered while keeping the massive potential of nanotechnology alive.

The principle of lowering risk is common in the investing world. Often, a few large investors will join forces to start a hedge fund. The idea behind a hedge fund is to maximize profit while lowering risk.

Hedge funding requires a large sum of money, gathered from multiple investors, that is diversified over many different investment options. Hedge funds have tremendous potential to improve the current investment outlook in nanotechnology. To successfully incorporate the use of hedge funds into the nanotechnology world, a determination of which nanotechnology ideas are the right ones to invest in must be made. Thorough research into the individual company along with a complete and in-depth review of the company's finances, management, research, and ideas is necessary. Once a company has been researched and is found to show promise, hedge fund investors must be brought on board.

Showing them the potential gain of a correctly chosen nanotechnology company will likely intrigue investors who will want to learn more. After weighing the risk with the reward, smart investors will inevitably invest in this type of well-researched and promising hedge fund.

The decision to combine the ideas of nanotechnology with hedge funding stemmed from the failure of the other ideas that nanotechnologists have used so far. Nanotechnology is relatively new. This only means that once a good idea to fund nanotechnology is found, it could become the most lucrative investment option for any type of investor.

First, it must be distinguished which methods are feasible. There are many different people who are willing to invest in small companies. However, they will want a large piece of equity in a potential company that they seek to invest in. Many times, the ideologists of nanotechnology are weeded out of the very companies they are creating. Due to the high costs of creating these genius ideas, many are faced with the harsh reality of having sunk their entire life savings into their creations. These same brilliant minds often end up with no better option than to sell their idea for a very small percentage of what it is actually worth. Nanotechnology has used most every type of investment idea, including venture capitalists, angel investors, joint ventures, and mutual funds.

The "valley of death," the patent issues, the lack of funding, and the relatively new technology, nanotechnology, are all related to one main thing: trying to bridge the gap between the research lab and the commercialization of the products. Nobody will invest because nobody else has proven that investing in nanotechnology is a wise decision.

There is history of a hedge fund, which was created purely for nanotechnology in the United States. In 2006, Global Crown Capital launched what it believed to be the first nanotechnology hedge fund and venture fund in the United States. The First Nanotechnology Fund (FNF) invests primarily in public nanotechnology companies worldwide, across a range of industries and economic sectors, in order to maximize potential returns and reduce volatility. The industries include healthcare, drug development, defense, chemical, and electronics, as well as others involved in nanotechnology. The fund deploys a long/short strategy, with trading supported by fundamental research and technical analysis developed by Global Crown Capital's analyst team. Rani Jarkas, managing director and chief investment officer of Global Crown Capital, said, "Most investors are aware that nanotechnology poses a significant investment opportunity, but they don't know how to evaluate it or how to get involved" (HFM Week).

"That combined with the fact that there is little credible investment advice available on the street or few suitable products makes this a very challenging arena for professional investors" (HFM Week).

FNF also invests in private, commercially promising nanotechnology companies with proven management teams, innovative and proprietary technologies, and the potential for significant medium-term returns. Its launch expands Global Crown Capital's presence in the nanotechnology investment community. The firm currently provides systematic, proprietary coverage of nanotechnology public equities worldwide and has developed a pure-play nanotechnology index to be used as a benchmark by professional investors (HFN Week).

More recently, hedge fund investor Nanostart AG is making preparations for ADR trading in the United States. Bank of New York Mellon (BNYM), one of the world's largest depositary banks, is providing active support during the process. Entry into the U.S. capital market will accommodate growing interest in Nanostart on Wall Street (Akesson).

The nanotechnology industry is hoping that this will be the first step in creating a positive buzz among the nanotechnology industry, speaking in terms of successful investing.

"We look forward to helping Nanostart unlock the investment potential of the U.S. investor community," said Michael Cole-Fontayn, chief executive officer of BNY Mellon's Depositary Receipts business. "Senior executives at Nanostart have been highly collaborative in their approach to accessing the U.S. capital markets, and we believe this will translate well to potential investors. As the world's leading depositary, BNY Mellon will utilize its many resources to develop a mutually beneficial partnership between our respective organizations" (Akesson).

ADRs (American depositary receipts) are certificates issued by a U.S. bank that securitize ownership of stocks in a company. ADRs are traded as shares of the company in the United States. Depositary bank for the Nanostart ADRs is Bank of New York Mellon, which employs over 40,000 people in 36 countries and reports assets of nearly 1.0 billion U.S. dollars. The trading center for Nanostart ADRs is anticipated to be the OTCQX market (Akesson).

The young and emergent OTCQX market provides U.S. investors with a trading platform that allows convenient access to foreign companies and ensures transparent trading as well as assurance of comprehensive information. In addition, companies traded there are not subject to the high costs associated with a listing on other American markets. OTCQX is now the home of numerous international corporations, including many DAX-listed companies such as Adidas, Allianz, and BASF (Akesson).

The Nanostart ADR program is described as a sponsored ADR Level 1 program. Ten Nanostart ADRs correspond to one Nanostart share. ADRs are denominated in U.S. dollars. With an annual volume of approximately 18 billion dollars, the U.S. capital market is by far the largest capital market in the world (Akesson).

Nanostart Investments has managed to achieve their capital objectives for 2010 in the first half of the year. The company also obtained a gross cash influx of nearly 1.3 million Euros ($1.75 million) through a capital increase in April 2010.

Nanotechnology has recently experienced a stage of exponential growth. The National Science Foundation has forecasted that $1 trillion in nanotechnology-enabled products will be on the world market by 2015. CEO Marco Beckmann manages the company's global portfolio and business activities with his investment team (Akesson).

The company's supervisory board, which consists of figures from the worlds of business and technology, actively participates in all-important decisions. Nanostart AG has a branch office in Berlin, and in 2008 it established a wholly owned subsidiary in Singapore, Nanostart Asia Pte Ltd. Nanostart has already sold eight portfolio companies, more than any other venture capital firm investing in nanotechnology (Akesson).

Alfred Jones, an American magazine reporter and former sociologist born in Australia, started the first hedge fund in 1949. As a staff writer at *Fortune* magazine, he wrote a story, "Fashions in Forecasting," that inspired him to try the stock market himself. With $100,000 to work with, he and four friends devised a two-part investment strategy. First, they borrowed additional funds so that they could invest far more than they had. Known as leveraging, it's the same technique used to finance the takeover of companies and is not all that different from borrowing money to buy a house. The true innovation was the second part of Jones's strategy. He picked not only stocks he expected to increase in value, but also stocks he expected to decline. He sold the expected losers "short," meaning he agreed to sell them on a future date at the current market price. So if the stock dropped, he could buy it at the lower future price, sell it to the buyer at the current price as agreed, and profit from the difference. His contract to sell the stock short meant that he had to buy at the future price even if he turned out to be wrong and the stock rose. But by betting that some stocks would rise and others fall, Jones "hedged" his position—hence the term "hedge fund"—reducing the effect that a sudden change in the general stock market would have on his investments. He made a lot of money being right more often than being wrong. Over the past decade hedge funds have roared back. The prospect of earning hundreds of millions of dollars in fees each year lured top investment bankers from Wall Street to hedge funds in Greenwich, Connecticut, and Palm Beach, Florida. At the same time, financial windfalls from a booming economy left thousands of the wealthiest Americans looking for places to put their money. The bankers and the wealthy met initially in hedge fund bliss. The numbers tell the story. From about 300 in 1990, the number of hedge funds rose to nearly 6,000 in 2001 and is currently estimated at 8,000 or 9,000. Assets under management in hedge funds increased from $39 billion in 1990 to $550 billion in 2001 to almost double that today. They are not regulated. To oversimplify slightly, a hedge fund is like a mutual fund that has been designed to avoid four federal laws that generally require investment funds and their advisers to identify fund officers and holdings and to submit to Securities and Exchange Commission oversight (Skeel).

The first law is the Investment Company Act of 1940. The 1940 act does not apply if the hedge fund has fewer than 101 investors. In 1996, Congress added a second exemption, waiving off the 1940 act if the investors are "qualified purchasers." A qualified purchaser needs $5 million if he is an individual and $25 million if it is an institution, but a hedge fund can theoretically have as many qualified purchasers as it wants. (In reality, the maximum is 499, for reasons that will become clear in a moment.) The thinking behind the exemptions is that a few friends, family members, or wealthy investors who are financially sophisticated don't need the regulatory protections of the 1940 act, safeguards like restrictions on risky investments, and limits on performance-based fees. Most funds choose the second exemption, because it allows them to tap a much larger group of investors (Skeel).

Hedge funds can avoid the next two laws, the Securities Act of 1933 and the Securities Exchange Act of 1934, by staying private and small. The extensive disclosure requirements of the 1933 act kick in only if the fund seeks money from the general public rather than from investors who are "accredited," meaning each has more than $1 million in net assets or earns more than $200,000 a year. And under the 1934 act, the fund doesn't have to file regular disclosure statements like quarterly financials if it isn't listed on a stock exchange and if it has fewer than 500 investors (which explains the limit of 499 qualified investors described above). Again, the thinking is that a few well-to-do friends and family members don't need the protections of the securities laws (Skeel).

The fourth law, the Investment Advisers Act of 1940, requires fund managers to tell the SEC how they're investing and to follow the commission's rules. But the act applies only if the manager has more than 14 clients and promotes himself to the public as an investment adviser. The trick here is in the definition of "client." The SEC says the term can mean an entire investment fund. Rather than counting investors, in other words, the commission will treat the hedge fund itself as the client (Skeel).

Avoiding these four laws allows most hedge funds to operate in secrecy and, unburdened by investment restrictions or the cost of public disclosure, do almost anything they want. Their options include hedging, the practice that Alfred Winslow Jones pioneered, but many funds don't hedge. Some focus on major corporate deals like the Mylan-King transaction and make money by predicting a deal's effect on a company's stock. Others use mathematical models to exploit blips in the prices of stocks, bonds, or other securities. Still others buy "distressed debt"—bonds and other obligations of companies in financial trouble. The funds pay a small fraction of the debt's face value, betting it will be worth more if the company's fortunes turn (Skeel).

A large concern lies within the hedge fund idea. Hedge funds are very complex and lucrative. They are low risk-high reward investments. Many investors do not invest in hedge funds because the transaction fees are too expensive. In the case of nanotechnology, they would be diversifying the investment into different nanotechnology companies and ideas. There are fees that go along with this. The investors pay for the hedge fund manager's salary. A typical hedge fund manager makes approximately 20 percent of the total profit of the fund. After fees and other costs, the investor will only see about 65–70 percent of the profit. This sounds like a lot but with the general lack of risk of hedge funds, it is much more lucrative than investing in other things. Today, certificates of deposit are returning less than two percent on average. The concern with hedge funds lies within the legality. Many investors may be wary of the legal issues and not want to invest regardless of the profits. Hedge fund law was created to allow the smarter investors to diversify their money in a very lucrative way. The system was not created for people to break the law. It was merely created like that to maximize the profits for hedge fund managers who were some of the smartest and innovative people in the world. They did not want to punish hedge fund managers by handcuffing their options and making it impossible for them to make money. Hedge funds are extremely low risk and very high reward, if managed correctly.

The hedge fund seems like the smartest way for nanotechnology to bridge the gap between research lab and commercialization. Venture capitalists could not bridge the gap from the research lab to a marketable product. Angel investors proved to be just a temporary fix and end up costing more time and money in the long run. The recent success of hedge funding in the field of nanotechnology, combined with the advancement of the scientific ideologies in nanotechnology, can only help nanotechnology succeed more quickly and effectively. Investors are more willing to invest because there has been a very recent success. Over the last 15–20 years, nothing positive ever came out of investing in nanotechnology. Over the last 1–2 years, however, there has been a change. People are starting to realize the benefits and capabilities of nanotechnology and the possible monetary gain from investing in early-stage nanotechnology companies.

The facts demand the conclusion that hedge funding will most definitely work in the field of nanotechnology. Ideally, early-stage companies will need a very competent hedge fund manager or board of directors. The primary focus of the job will lie in getting the correct nanotechnology companies on board. Determining if a particular nanotechnology company is potentially great is difficult. The person in charge of determining this must do his or her due diligence. The company must be examined thoroughly. Trust is an important factor for each person involved in making the nanotechnology aspect of the company successful. The idea of the nanotechnology must be an innovative and marketable one. There must be a hedge fund manager who can diversify the investments correctly. He must allocate the money into the correct areas. As long as there are a small number of large investors and a few genius nanotechnology ideas, there will be more profit for each person involved than they could ever hope to receive. The potential is enormous. The idea is intriguing and would lure most savvy investors to think very hard about joining.

Nanotechnology will be a huge market in 20–30 years. This idea speeds up the process, making everyone who was willing to take a small risk very wealthy. They will be the first to successfully market and sell nanotechnology products. Hedge funding will take nanotechnology through the "valley of death" and into the promised land.

9

Business Applications

The best advice for anyone taking his or her product commercial is: Get a good accountant. Get a good attorney. Get a good business coach. Get a good marriage counselor. You are going to need all of them at one point or another. The effort you invest in making your dream come true will make you wise beyond your years in a short time.

I can remember when I began my doctorate in industrial/organizational psychology. I had just become a COO of a technical company a few months before. I can remember getting down on my knees and praying for God to allow me to learn the lessons I needed to learn in business so that I could make a difference with my education to help businesses that were having a difficult time realizing their dreams. Needless to say, God took me seriously; I have paid my dues. I learned to sweat blood and tears trying to create payroll. I achieved the unique ability to swallow my tongue and my pride in difficult, but meaningful negotiations. I added depths of vision, seeing beyond what appears in front of your face. Paradigms are like diversity uniforms that I can wear for an occasion so as to understand my fellow man and not lose my own. Having my own business has been an experience I would not give up. For the weak of spirit, I would say run; run until you think your heart will explode and never look back. Get a cushy job and prop your feet on someone else's desk and collect your paycheck from the sweat from their brow. For those of you committed to making your business dream come true, roll up your sleeves and dive into the greatest and most rewarding experience known to mankind.

Now to pull all of this information into a working model for your nanotechnology dream.

In the early chapters of this book, you learned about the types of nanotechnology business out there and some basics on how to create a business plan to focus your company toward success as well as to gain the necessary investors to help through the difficult financial times from inception to manufacturing. You learned of the dangers, the ethics, and the support of multiple centers and organizations out there to help you with difficulties along the way. This chapter is going to take you through some of the basic business applications decisions you need to make to get your company started on a solid foundation.

Accounting

Finding a good accountant that knows the ins and outs of business as well as the intricacies of your industry and that actually can deal with your personality will take time, but is well worth the energy invested. Not all accountants are created equal. Most have bookish personalities that love to pore over numbers and take your money very seriously, sometimes giving you advice for which you want to fire them. Hear them out—they are trying to save you and your business (Figure 9.1).

Find an accountant that will stand up to you. It is uncomfortable, but businesses can be ruined by well-meaning accountants that do not want to stress a boss who is under mountains of undue stress already, and will tarry on giving you the bad information. That is the kiss of death. You are a business person, responsible for the livelihood of your employees and for the financial success of your company. You have been duly warned that it is not a job for the meek. If you can't stand the heat, hire a business manager. They thrive on chaos, stress, and last-minute saves.

In my company, I do business turnaround consulting for companies having difficulties. I go in with the mission of any good business doctor to stop the corporate bleeding. Quite often it is the cash flow. One particular company with whom I was familiar had an accountant who was highly qualified and worked in retail for 30 years, had been with an up-and-coming

FIGURE 9.1
If your accountant is not giving you regular financials of the company to review, get rid of him and either hire an accounting firm that you can speak to openly and freely, or hire an accountant that has some fire for your business and does not allow for mistakes.

construction company for about a year. The owner had hired the accountant and never looked back, pushing forward to get new business and work on other parts of the busy firm that was contracting to universities, etc. When he needed money for a motorcycle trip, or to take a vacation, or even to purchase a heavy-duty extend cab, the money was always there. He loved the accountant, until he was contacted by the IRS and was on the verge of losing his company.

His comment upon contacting me was: They were busy, working on huge deals, how could there be money problems? One of the first things I did was to ask to see the books, which the owner replied he had not personally seen for over 6 months. Every time he asked, the accountant was factoring the numbers out further, learning his way from his retail chain past to a construction company scenario with large material purchases, buying on time, cash flow problems, etc. The accountant did not want to stress him and felt if he could break the numbers down further he would find a way. I opted to go in and clean house, firing the accountant and hiring a firm that gave weekly reports and could, within hours, provide any financial statement the owner needed for any purpose. Why? Because multiple hands of a firm understand the industry and cash flow needs.

Know where your money is, make friends with it, do not fear it or abuse it. Most people do not have sound financial training. Take some classes. Read some books. Do not rely on someone else to handle your money successes or problems; keep your hand in it. Know what your trusted advisors are doing. You don't have to know more, but you do need to know how to read financial reports (at least weekly) and know enough to guide them, or to pick up on a problem. Embezzling happens because people are not careful, trust too much, and do not have enough basic financial information to protect themselves.

Sales/Marketing

You can be the next Einstein and have the greatest business idea and plan in the world, but if you don't have customers, you don't have a business. Money makes the world go round, and your sales and marketing team need to articulate your vision to the clients to sell your products.

Sales and marketing employees, just like accountants, need to be carefully screened and matched to your product and your company. The best face in the application pool that has a gift of gab and "suck up" may sell you on themselves, but can they sell your company and your product?

CASE STUDY: SALES AND MARKETING

I recently worked with a company that hired an amazing young salesman who was a great orator and could speak to crowds and anyone that approached him. His background had been in retail sales—shoe sales. He was the best in the business and was ready to break into the big bucks that sales offered and applied at a service company that immediately hired him. After the initial training, the brilliant smile and charisma turned to sullen depression as the frustrated salesman grudgingly used the phone to make his appointments. His numbers went down. He had not made a sale after 6 months on the job. His personality was biting to others in the office. Thinking that he just needed to go out more and not work the phones to contact clients for appointments, the owner put him out on the streets to make cold calls. The salesman came back into the office more depressed, more disgruntled, and hated sales. He was frustrated that people stared and often glared at him when he walked into businesses.

I was contacted by the company to try to salvage the young man before he was terminated, as he had already cost the company thousands with no return on their investment. The results—the salesman was a retail salesperson and did not like business-to-business sales. He hated phone work. He hated the paperwork. He hated cold calling and was paranoid when people looked at him strangely even though he had a shaved head, wore faux 2-carat diamond earrings, and had tattoos. In the retail world of shoes, he was the cool, charming guy that was a little chancy, but could charm any tired foot into a shoe. He could sell when people came up to him and told him what they wanted.

Conclusion: Make sure the personalities fit your product. The best salesperson in the world just may not fit your line and can cost you more than he or she brings in. Also, the pay scale is often factored on the industry. Most are expected to bring in two to five times their salary or more, depending on the industry. Do your homework before hiring. Will it be salary, or a draw against commissions, straight commission, or a stipend and commission?

HR/Talent Development

Nanotechnology brings with it special problems in hiring even basic people, as many countries do not have the technical background and training needed for the job. Even positions such as administrative assistants need to know the terminology and basics of the industry. Universities and colleges are just beginning to graduate qualified candidates.

Getting your HR department and support staff up to the level that they can be productive can take years without training. Training companies are just beginning to look at this problem and groom the talent for new companies.

For those companies lucky enough to find good employees with the technical knowledge and training needed, it can be a task to keep them, as there is a shortage at this point for this type of worker. Meeting the needs of the worker with regards to salary, additional education, and coaching is a key investment into the workers that often keeps them loyal. The technical "toys," often available in the technical corridors, along with the likemindedness of the group and support system, are attractive extras for this type of worker.

Workers from other countries such as India or China have strong technical backgrounds, but may find it difficult to work as expatriates because of the new antiterror policies that do not allow foreign nationals in many of the labs, both university and government.

Continuous Improvement

While the confines of this book are too narrow to give all the business commercialization secrets, one that I would challenge all company leaders and key development people to do is to become enrolled in a program called Lean Six Sigma. I will briefly look at the program designed to move your company toward efficiency and excellence.

Lean Six Sigma is a management methodology that is driven by data and integrated into your business improvement process. The central idea is that if you can measure how many defects you have in a process, you can systematically determine how to eliminate them to approach what is called zero defects. It is about reducing variation and finding out the facts about your business process before acting, thereby creating better product, increasing efficiency, and enlarging your profit margin.

A Six Sigma initiative is implemented through five major stages and begins with what is known as a deployment program from the top down, so that all understand the concept and it permeates the company. Individuals go through the required trainings to become certified belts. The program initializes Six Sigma in the company by establishing the goals and installing the infrastructure of the company to support it. By assigning, training, and equipping the staff with the information and tools, it is possible to expand the scope of the original initiative to additional organizational units. The program is kept strong and fresh via constant realignment, retraining, and actual evolution of the company. It is a great way to build a core unit, and to build a company team.

Insurance

Earlier in this text we read about Lloyd's of London and its report on the emerging risks of nanotechnology. Its concerns were (1) nanoparticles—the fact that the particles can be much more reactive than larger volumes of the same substance and the instability that could become volatile and extremely toxic with manufacturing mistakes; (2) impacts on health—whether from the nanoparticles themselves as toxic agents or the possibility of buildup within the body and the subsequent toxicological effects; (3) unknown impacts on the environment (ecotoxicology)—the possibility of the particles being absorbed and moving up the food chain, or the effect of some nanoparticles (copper or silver) that can be harmful to aquatic life; (4) positive effects of nanotechnology—the seductive charm of money and what nanotechnology can do contributes to the rush to capitalize and commercialize before assessing the safety of the product; and (5) lack of regulation—the lack of specific regulation and standardization can open the door for additional litigation.

Knowing where your product fits in this emerging risk report is important in protecting yourself as a business owner when pursuing insurance. As stated many times in this text, there is a lot the common layperson does not know about the intricacies, dangers, and miraculous discoveries made with nanotechnology. It is the inherent responsibility of the business owner to make sure that he or she acquires an insurance firm or company with knowledge to protect him or her and share resources with his or her company. Ignorance, in this case, is not bliss.

Legalities

There are always cautious people and naysayers in the wings touting the dangers of growth of new technology, and they are often brushed aside until there is a major catastrophe, such as the meltdown of the nuclear plant in Japan. We need both sides, the naysayers and those who push to make discoveries a reality. But there is a fine line of caution that everyone needs to follow with the complexities of working on a molecular level. The possibility and inevitability of an accident needs to be discussed. Laws for standardization need to be in place. Protection systems for national security need to be reviewed in how to handle the new technology. Laws need to be written.

James Brindell (2009), in a recent article for the *Florida Bar Journal*, stated:

> The nanotechnology industries should consider encouraging federal pre-emption and the placing of regulatory control within one federal

agency in exchange for mandatory reporting of the development and use of nanomaterials, the development and disclosure of test data, and the labeling of consumer products. Focusing the data collection and regulation within one federal agency would preclude subjecting the nanomaterials industry to potentially overlapping and conflicting regulations, would minimize the possibility that an issue of concern might fall unattended between agency programs, would enhance the public's confidence that their health and safety are being protected, would produce more comprehensive and fact-based public policy, and would reduce the likelihood that the government would be an unnecessary obstacle to the advancement of nanotechnology.

Advice: Keep your money dry and create your exit plan as you create your business plan.

CASE STUDY: FUNDINGPOST

For more than 10 years FundingPost has been connecting thousands of angel and venture capital investors with entrepreneurs, representing over $106.8 billion. FundingPost has hosted 185+ sold-out venture events in 20 cities over the past 9 years.

> With over 10,000 CEOs and 700 Venture Capital Funds attending events in 20 cities nationwide; a Printed Dealflow Magazine; and a deal-exchange website with over 7,700 VC & Angel Investor members and over 133,000 companies, that has, on average, made an introduction of an Investor to an Entrepreneur every business day since its inception; FundingPost believes that it is important to reach investors in every medium possible—both online and offline. FundingPost has been responsible for millions and millions of dollars in venture capital raised!

Joe Rubin is the director of Fundingpost and sends weekly/monthly updates to those interested in meeting with hundreds of like-minded and serious businessmen and -women and investors at each conference. Joe and the staff run a heavy schedule to meet the needs of their clients, but each conference shares a focus of those attending and everyone comes away a winner. The group promises: "There will be plenty of time for networking with the Investor panelists, both before the panel and after the panel at the cocktail party!"

The following is a list of upcoming conferences for the 2011 year. Additional conferences and locations are posted on the website at wwwfundingpost.com when they are scheduled.

June 16, 2011—Los Angeles, California: SoCal Angel Investor and Early-Stage VC Conference. Early-stage venture and angel conference with a pitch-coaching workshop. Meet Early-stage VC funds and angel groups in Southern CA. Meet angel and VC speakers from Tenex Medical Investors, HOT Ventures, Ventana Capital, and DFJ Frontier, with more TBA. Sponsored by Bingham McCutchen.

June 23, 2011—Silicon Valley, California: Silicon Valley Angel Investor and Early-Stage VC Conference. Early-stage venture and angel conference with a pitch-coaching workshop. Meet early-stage VC funds and angel groups in Northern CA. Meet investors from Qualcomm Ventures, Nokia Ventures, Rustic Canyon Venture Partners, Launch Capital, the Angels Forum, and more TBA. Sponsored by Bingham McCutchen.

July 21, 2011—Washington, D.C.: Washington D.C. Angel Investor and Early-Stage VC Conference. Early-stage venture and angel conference with a pitch-coaching workshop. Meet early-stage VC funds and angel groups in the DC area. Meet investors from Edison Venture Fund, Amplifier Venture Partners, and NAV Fund, with more TBA. Sponsored by Bingham McCutchen.

August 25, 2011—Chicago, Illinois: Chicago Angel Investor and Early-Stage VC Conference. Early-stage venture and angel conference with a pitch-coaching workshop. Meet early-stage VC funds and angel groups in Chicago. Angel investor and VC speakers TBA.

We will be discussing trends in early-stage investing, things that are most important to investors when they are considering an investment, the best and worst things an entrepreneur can do to get the investors' attention and grow the business, and additional advice for entrepreneurs. Sponsored by Perkins Coie.

Wednesday, September 7, 2011—New Haven, Connecticut: New Haven Angel and VC Event. Early-stage venture investing. How to meet investors, pitch to them, and what it really takes to get them to write you a check. We will be discussing trends in early-stage investing, hot sectors, sectors that these angels and VCs look at, things that are most important to them when they are considering an investment, the best and worst things an entrepreneur can do to get their attention, additional advice for entrepreneurs, and of course, the best ways to reach these and other investors. Sponsored by Connecticut Innovations.

Thursday, October 27, 2011—Miami, Florida: Media and Entertainment Investing Conference. Interested in meeting early-stage investors in the media and entertainment sectors? If you are an investor, company, service provider, or member of the press in music, television, video games, film, design, or animation, either on the creative side or the technology that supports it, this is a must-attend event! Sponsored by the Launch Pad at the University of Miami.

The panels for the day:

- Early-stage media investing: How to meet investors, pitch them, and what it really takes to get them to write you a check.
- Growing your media entertainment business: Protecting and selling your intellectual property.

See one of the URLs below to register for these events or to peruse upcoming events:

http://www.fundingpost.com/events.asp
http://www.FundingPost.com
http://www.twitter.com/fundingpost
http://www.facebook.com/fundingpost

FundingPost/Second Venture Corporation
7365 Main Street, Suite 324
Stratford CT 06614
800-461-5509

10

Support Organization

The support of individuals and organizations to the nanotechnology commercialization effort is invaluable as it takes into account the years it can take off going to market as well as the emotional business support that owners have in securing funds, patents, locations, and employees to make their enterprise effective and profitable.

We will visit various groups that are dedicated to the growth and development of these businesses and the technology.

Nanotechnology Commercialization Group

The Nanotechnology Commercialization Group (NCG) is a single storehouse for university nanotechnology intellectual property and grants a valuable marketing advantage. The NCG began as an agreement between Drexel University, the University of Pennsylvania, and BFTP/SEP and operates as an administrative unit of the University of Pennsylvania's Commercial Development Office.

The NCG, staffed with two Penn employees and one Drexel employee, each with nanotechnology expertise, is located in the commercial development office.

NCG Administration

Anthony Green, PhD
Ben Franklin Director, NTI
Vice President of Technology Commercialization:
  Life Sciences, BFTP/SEP
anthony@sep.benfranklin.org
(215) 972-6700 x3713

Erli Chen, PhD
University of Pennsylvania
Director, Nanotechnology
  Commercialization Group
chen@ctt.upenn.edu
(215) 898-9272

Philip Caldwell
Associate Director, Technology Licensing
Representative, Nanotechnology
  Commercialization Group
Drexel University, Technology Commercialization
pcaldwell@drexel.edu
(215) 895-0999

Shilpa Bhansali
Assistant Director, Nanotechnology
  Licensing
Center for Technology Transfer
University of Pennsylvania
sbhansali@ctt.upenn.edu
(215) 573-4307

Vijay Iyer, PhD
Licensing Associate
Office of Technology Transfer
Temple University
vijay.iyer@temple.edu
(215) 204-7619

Greg Baker, PhD
Associate Director, Commercialization
Technology Transfer Office
Children's Hospital of Philadelphia
bakerg@email.chop.edu
(215) 590-5645

The NCG, together with support from BFTP/SEP, provides the following services for all university members of the NTI:

- Evaluating commercial potential
- Developing commercialization strategies
- Marketing
- Negotiating licenses
- Facilitating the formation of start-up companies

The NCG is the ideal starting place for investors and companies searching for licensable technology. It is also the perfect place for university researchers to begin their journey toward the commercialization of their research.

Foresight Institute: Studying Transformative Technologies

The Foresight Institute was founded in 1986 as the first organization to educate society about the benefits, dangers, and risks of nanotechnology, Artificial Intelligence (AI), biotech, and similar life-changing technologies. The Institute is a leading think tank and public interest organization that focuses on transformative future technologies. Accurate and timely, Foresight provides balanced information to help society understand nanotechnology through its publications, public policy activities, and conferences.

The converging technological capabilities ranging from MEMS and NEMS to molecular biology are developing at a steady pace and "have the ability to change the world in radical ways in the coming decades—radical, but not unforeseeable." By using all the intellectual tools available to help understand the changes that the new technologies are going to affect humans with, the Foresight Institute studies the technology via history to computer modeling, and AI to garner information to help the human condition and educate the public.

The Foresight Institute is a member-supported organization of corporations and individuals that are interested in making sure that the future of nanotechnology "unfolds for the benefit of all." The membership currently includes over 14,000 concerned individuals and a growing body of corporations that include scientists, engineers, business people, investors, publishers, artists, ethicists, policy makers, interested laypersons, and students from grammar school to graduate level. Foresight is a 501c3 nonprofit organization. Donations are tax deductible in the United States. Contact: Christine Peterson, telephone +1 (650) 289-0860, ext. 255, peterson@foresight.org.

The Organization for Economic Cooperation and Development (OECD)

The Organization for Economic Cooperation and Development (OECD) is an international organization. The OECD recently released a Nano Risk Framework for risk managers. The EU is embracing the approach, while the United States and Japan prefer a lighter regulation.

The Center for Responsible Nanotechnology™ (CRN)

The Center for Responsible Nanotechnology (CRN) is a nonprofit organization, founded in 2002. The vision of CRN "is a world in which nanotechnology is widely used for productive and beneficial purposes, and where malicious uses are limited by effective administration of the technology."

CRN is the brainchild of Mike Tredor, executive director, and Chris Phoenix, director of research. Tredor has a background in technology and communications company management. He is a member of the executive advisory team for Extrophy Institute and serves on the boards of directors for Human Futures Institute and World Transhumanist Association. Phoenix has a master's of science in computer science from Stanford University. He is an inventor, entrepreneur, and published author in the fields of administration of nanotechnology, nanomedicine, and nanomanufacturing.

CRN was formed to advance the safe use of nanotechnology. With the devastation of the Japanese earthquake, tsunami, and subsequent radiation emission, the information is changing as life is pulled back together. Information for some time will be forthcoming on the changes in stats and progression.

The following is a report from the university days after the earthquake. Tohoku website, http://www.tohoku.ac.jp/english/, April 7, 2011

MESSAGE FROM THE PRESIDENT:
3RD MESSAGE: MARCH 25, 2011

I would like to offer my deepest thanks to all of you who have supported and encouraged us following the Tohoku-Pacific Ocean Earthquake. I would also like to offer my condolences to all of you who are suffering tremendous fatigue due to the hard day-to-day conditions.

Two weeks have passed since we were hit by the earthquake, which was of a historically unprecedented scale. Gradually, Tohoku University is being restored, and is preparing to make great strides forward.

Fortunately, we have ensured safety and have suffered no human losses on the Tohoku University campus. This disaster did not stop our university's activities for society; immediately following the earthquake, the Tohoku University Hospital operated tirelessly to provide medical care and examinations in the affected areas, the Graduate School of Dentistry assisted in identification of individuals, and the Cyclotron and Radioisotope Center dedicated itself to monitoring radiation levels. I have also heard that our students have been active in volunteering, wanting to help those affected by the disaster in any way they can.

At times like this, our sorrow should be transformed into hope. We at Tohoku University will exert our collective efforts to contribute to the regional society. At the same time we will bring together our wisdom for the restoration and revitalization of the region. Finally, we will strategically and systematically address research that will lead Japan into a new era, and disseminate and apply our research findings.

While some of our university's facilities and equipment were damaged due to the earthquake, I believe we are now in a situation to further realize Tohoku University's functions and abilities. In addition to immediate and complete restoration of our research and educational infrastructure, we will move forward in order to realize even greater strides in our educational and research capabilities, and ability to contribute to society. Once again, I ask you all for your cooperation in this matter.

Akihisa Inoue
President, Tohoku University

Website: http://www.wpi-aimr.tohoku.ac.jp/tmp/index_en.html

DIRECTOR'S MESSAGE TO AIMR
RESEARCHERS (REVISED MARCH 29)

AIMR is now restoring our own research activities from day to day. I hope to meet all researchers at AIMR and discuss the necessary measures and the future orientation of research as soon as possible. Thank you for your attention and cooperation.

AIMR researchers' current situation is as follows.

Researchers' current situation (as of April 4) (This information will be updated regularly.)

For your information, the period employees do not arrive for work due to the disaster is treated in the following manner at University: Employees who do not arrive for work are required to take their own paid leave except the case;

1. authorized leave mentioned in the official notice by the Head of the Disaster Headquarters issued on March 16 (see below a tentative translation), or
2. special leave for employees who do not have adequate commute method. In this case, full salary will be paid.

However, if you do not come to work without above-mentioned leave, it will be treated as absence from work without pay and your salary will be reduced.

An official notice on the temporary measures about the work issued on March 16. (PDF file)

Please see also the official site of Tohoku University.

http://www.tohoku.ac.jp/english/

For more information on the disaster in foreign language, you can contact Sendai Disaster Multilingual Support Center Sendai International Relations Association.

Tel 022-265-2471 or 022-224-1919
Fax 022-265-2472
http://www.sira.or.jp/saigai/

OTHER INFORMATION

- Tohoku University Earthquake Disaster Relief Donations (International Remittance)
- We would like to extend our deepest gratitude for your kind consideration to WPI-AIMR and Tohoku University following the Tohoku-Pacific Ocean Earthquake of March 11. In order to handle this situation, Tohoku University Earthquake Disaster Relief Donations was established. Please refer the following URL for details. http://www.tohoku.ac.jp/english/contributions.html

  Your generous donations will be used to restore the research infrastructure and educational environment for students at Tohoku University. If you would like to contribute to restore WPI-AIMR especially, please indicate so on the form.
- About recovery of the lifeline
  - WPI network in Katahira area has been recovered.
  - Electric power and water supply in Katahira area have also been available. Supply of gas is not planned at the moment.

Yoshinori Yamamoto
WPI Center Director, Tohoku University

http://www.tohoku.ac.jp/english/contributions.html

How to Make Tohoku University Earthquake Disaster Relief Donations (International Remittance)

We would like to extend our deepest gratitude for your kind consideration following the Tohoku-Pacific Ocean Earthquake of March 11.

We would like to inform you of how to make contributions. Your generous donations will be used to restore the research infrastructure and educational environment for our students.

Unfortunately, we are unable to accept goods (ex: food, water, clothes etc.)

- If you wish to donate to Tohoku University Hospital, please click here (in Japanese):
- If you would like to donate from inside Japan, please follow the Japanese instructions.

Please send the following information to our contact e-mail (sinsai*bureau.tohoku.ac.jp [Please replace "*" with "@"]) and we will provide our bank account information by return e-mail. We cordially ask you to send your remittance to the bank account mentioned in the e-mail. (We apologize for the inconvenience, but we can only respond to English e-mail)

- Amount of Donation
- Surname, First Name
- Name of Institution, Job Title (optional)
- Postal Address, Postal Code, Country Name
- Telephone Number
- Fax Number
- Would you permit us to print your information in our public relations materials? (Please select one from a to d)
  a. Name, Name of Institution & Job Title, Amount of Donation
  b. Name, Name of Institution & Job Title
  c. Amount of Donation
  d. Other (Please specify)

International: Yayoi Kobayashi (Ms.)
Domestic: Takashi Kondo (Mr.), Toshiyuki Hachiya (Mr.)
Office of Tohoku University Earthquake Disaster Relief Donation, Tohoku University
TEL: +81-22-217-5578 / FAX: +81-22-217-4846
E-mail: sinsai*bureau.tohoku.ac.jp (Please replace "*" with"@")

[TOHOKU UNIVERSITY'S NEW CHALLENGES BRAND—NEW TOHOKU UNIVERSITY]

Faced with an almost unprecedented disaster, we do not aim merely to reconstruct and restore the region, but rather seek to implement a paradigm shift to a new human society, based on "Constructing a Safe and Dependable Society."

Our facilities and equipment suffered some damage in the disaster, but our functionality and true worth has not been reduced.

As a center for academic research and regional regeneration located in the disaster area, Tohoku University pledges to support the reconstruction and regeneration of local communities, both through the individual efforts of specialists and through systematic initiatives by the university as a whole by transmitting and putting research results into practice.

Finally, for the sake of future generations we will push forwards with efforts to create a safe and dependable society living in harmony with nature. Through these efforts Tohoku University will endeavor to support the evolution of human society while becoming a university worthy of the trust, respect, and affection of the world.

Tohoku University Project for Disaster Restoration & Regeneration of Afflicted Areas

---

Examples of Projects

Developing a society for future generations that can live in harmony with nature!

☐ Building a Safe, Secure & Harmonious City
☐ Handling of Earthquakes and Tsunami
☐ Handling the Nuclear Power Risk
☐ Regeneration of Local Primary Industries
☐ Regeneration of Local High-order Industries
☐ Regeneration of Innovative Industry
☐ Medical/Health Policies
☐ Restructuring of Lifeline Utilities
☐ Upgrading of IT Infrastructure
☐ Energy Policy
☐ Culture/Arts Conservation
☐ Human Resource Development & Utilization
☐ Enactment of Laws to Support Disaster Victims and Restoration

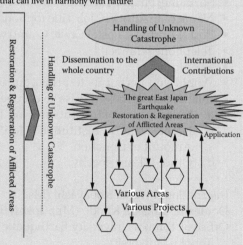

April 2011
Akihisa Inoue
President of Tohoku University

http://www.tohoku.ac.jp/english/2011/05/news20110512-01.html

We also thank you very much for your offering kind and warm contribution to the "Tohoku University Earthquake Disaster Relief Donations."
We cordially ask you to send your donation to the bank account below.

Name of Bank: The 77 Bank, Ltd
SWIFT Address: BOSSJPJT
Branch Name: Head Office
Branch Address: 3-20, Chuo 3 Chome, Aobaku, Sendai, Miyagi, Japan
Beneficiary's Account number: 100-7921144
Beneficiary's Name:
Kokuritsu Daigakuhoujinn Tohokudaigaku(shinsai-kifu)
Beneficiary's Address:
1-1, Katahira 2 Chome, Aobaku, Sendai, Miyagi, Japan

For our information, could you please send the following information by e-mail when you donate?
(to:sinsai@bureau.tohoku.ac.jp)

- Surname, First Name
- Name of Institution, Job Title
- Postal Address, Postal Code, Country Name
- Telephone Number
- Fax Number
- Amount of Donation
- Would you permit us to print your information in our public relations materials?
- (Please select one from a to d)
    a. Name, Name of Institution & Job Title, Amount of Donation
    b. Name, Name of Institution & Job Title
    c. Amount of Donation
    d. Other (Please specify)
- Notes

Thank you again for your thoughtful generosity.

CASE STUDY: THE NANOTECHNOLOGY GROUP, INC.: A GLOBAL EDUCATION CONSORTIUM

Contributed by TNTG.org

The nanotech revolution waits for no man, woman … or child.

Judith Light Feather is author of *Nanoscience Education, Workforce Training and K–12 Resources*, the first book of its kind that offers a practical strategy to bring awareness of nanotechnology to both public and private sectors of every level, starting at grades K–12, "the effective gestation point for future ideas and information."

IT ALL STARTED WITH A VISION

Judith Light Feather, the founder and president of the NanoTechnology Group, Inc., an organization focused on nanoeducation and workforce training, relates that it all started with a vision of a fully functioning complex interactive web of communication with computers that allowed students anywhere in the world to connect and explore ideas together from their classrooms. The year was 1995, and little did she know that the Internet would soon become that complex interactive web of that vision.

The purpose and mission of the NanoTechnology Group is to provide a supportive and collaborative advisory role to the members for facilitation and development of innovative nanoscale science education. This would include subject-specific math curriculum targeted for grades preK–20, featuring interactive virtual nanoscience curriculum and classrooms for global access and virtual interactive nanoscience laboratories (nanolabs) for experiential learning, along with massive multiplayer online role-playing games (MMORG) for education globally.

The goals, as a solution-based organization, also include the continuous postings of new resources for teachers, students, and parents on its websites, while seeding the ideas into the global matrix for widespread implementation. The organization was built on collaborative dialogues and strong professional relationships with members of like-mind who could contribute resources as a solution for global nanoscience education. It was never intended to be a commercial corporation that would profit on educational products; rather it forged ahead on a new path of global inclusion to freely provide nanoscience education to every child through Internet technology and virtual classrooms. "We aren't there yet, but we are determined to succeed in this decade."

History

The road began at NASA's Johnson Space Center (JSC) in 1995, where a journalist media pass allowed Light Feather the opportunity to explore the NASA educational learning resources for K–12 teachers, while learning more about the International Space Station and its missions. A year spent as an invited member of a think tank to discuss combining spirituality and science for the future of ethical space colonization and terra forming of planets which expanded her vision for the future of education. This led to a proposal for 21st-century education to connect classrooms to the International Space Station through the Internet in 1996, as a hub for students to interact globally from space to expand their earth-based perspectives with a virtual learning experience. Since only 14% of the schools nationwide were connected to the Internet in 1997, with limited computer labs, the project was a decade ahead of the technology.

However, this step led to requests from the National Learning Foundation in Washington, D.C., for Light Feather to join their efforts as director of project development and evaluate their agile skills program that would base all curriculum on attainment of specific skill levels, rather than rote memorization of disconnected facts. The focus on inclusion of new skills levels that would be necessary in the 21st-century workforce was still considered too futuristic for immediate funding. Those same skills were further developed by cognitive scientists and are now required as benchmarks for teachers to follow in each grade.

Nanoscience Enters the Arena in the Late 1990s

By now the year 2000 is fast approaching, and new discoveries of an infinitesimal size of science are surfacing that will soon astound scientists in every field with its amazing quantum property characteristics and the newly developed ability to move and manipulate atoms for commercialization. The Nanotechnology Coordination Office in Washington, D.C., directed by Dr. James Murday, became a member of the group at this time and has been an advocate and supporter of its educational vision since 1998. By 2000, the first National Nanotechnology Initiative was in place and universities began to receive funding for research and development of new courses integrating physics, chemistry, biology, and engineering to enable exploration of nanoscale science and the potential for future technologies that would change the world.

These developments also intrigued NASA Johnson Space Center (JSC) as they were proposing to fund nanoscale science research based on the late Richard E. Smalley's 1996 Nobel Prize in Chemistry for the discovery of C_{60}, a new class of carbon structure in the shape of a soccer ball, dubbed a buckyball or buckminsterfullerene. They were interested in how this discovery would be applicable for space applications, and Rice University also received funding for its Center for Biological and Environmental Nanotechnology (CBEN).

The excitement around this new size of science opened the door for Light Feather to become the assistant executive director for the NanoComputer Dream Team and develop its Nanoeducation Gold Team for exploration. The CBEN became its first university member and Director Kevin Ausman was excited about the prospects of developing nanoeducation for K–12. He was very helpful in providing information for the team along with the schedules of nanoscale science lectures at Rice University. In 2000, NASA held its first NanoSpace 2000 conference and the relevance of the potential for space applications of nanoscience became apparent.

Soon after the conference, the Gold Team was contacted by Milind Pimprikar, owner of the Center for Large Space Structures in Montreal, Canada, who wanted to develop a global space organization. Light Feather's advisory role in his project initiated the development of CANEUS—Canada, Europe, U.S. Space agencies—as a global consortium promoting micro/nanotechnology for space applications. By 2002, Pimprikar scheduled the first conference in Montreal, Canada, and CANEUS has now grown to include many of the new space agencies developing around the world, including Japan, China, India, Brazil, and Italy. Primprikar continued his association with the group and became a supportive member of the NanoTechnology Group when it was organized.

Birth of the NanoTechnology Group

The Gold Team membership grew so rapidly that by 2002, the NanoTechnology Group, Inc. (TNTG) was formed as a separate entity and incorporated as a 501(c)3 foundation for education. The application for the IRS designation took 1½ years to receive approval as the field of nanoscience was so new, and since TNTG did not charge a membership fee, it did not fit within the normal structure in education classifications. Determined not to give up, Light Feather submitted a multitude of emails requesting membership and the invitation to participate in the expert working group at the International Center for Science and High Technology (ICS)—United Nations Industrial Development Organization (UNIDO) *North-South Dialogue on Nanotechnology: Challenges and Opportunities*[*] in Trieste, Italy. The IRS was finally convinced that TNTG was a legitimate

---

[*] http://www.tntg.org/documents/52.html.

organization and should be approved as a foundation. Since TNTG was not receiving actual funding, the designation allowed the corporation many avenues of income, such as listing the value of volunteers' time along with in-kind donations, and donated expenses for travel and meetings that were important to their growth. This was a very important milestone as the IRS designation allowed partnering with collaborative universities to accomplish its goals as a global consortium.

News Division Provides Public Information and New Opportunities Globally

In the early stages of the development of the Global Consortium, a NanoNews Division* was established to inform the public about the potential and possibilities of the commercialization aspects resulting from the research and development of nanoscale science. Press releases were scarce in 1998–2000, but with the advent of the first U.S. National Nanotechnology Initiative, press releases quickly increased an avalanche of information. The potential was so great that the excitement spread across the globe and many countries allocated funding to their universities to teach nanoscience and discuss avenues that might include master's degrees in nanotechnology.

The NanoNews Division received invitations to participate in media tours of Switzerland in 2002, which included visits to all Swiss universities that were reorganizing their courses to integrate this amazing scale of science. The tour included visits to important commercial research laboratories such as IBM in Zurich, where the advanced scanning electron (SEM) and atomic force microscopes (AFM) were developed in 1995, hence quickly being acquired by universities as they allowed the viewing and manipulation of atoms.

The professors were as excited as children when they gave their presentations and encouraged the media participants to feel the surface of an atom with an AFM. The computer screen depicted the surface of the atom, and by placing a hand on the lever, one could actually feel the surface of the atom as the lever was moved.

A visit to the University of Basel to experience their virtual nanolaboratory online was also a unique experience. Discussions with the inventor of this technology reinforced Light Feather's vision that every child would eventually be able to experience learning with a virtual nanolab. Of course, bringing the project back to the United States and seeking a funding initiative to approve it turned out to be much more difficult.

* http://www.nanonews.tv.

Partnerships for Nanoscience Funding

Upon returning from Switzerland, Light Feather started the search for funding of a virtual nanolab by submitting a proposal to a Department of Education solicitation to develop new science curriculum. The proposal reviewers suggested that any nanoscience curriculum for K–12 would have to be funded by the NNI under the National Science Foundation (NSF).

The UVA Virtual Nanolab* finally received some funding by the University of Virginia in 2005, but has been underused and underfunded since then. Nanoscience is not included in the U.S. national or state standards for testing; therefore, teachers are unaware of the resources and are not required to use them.

Due to these issues, the NSF was able to have curriculum developed by the universities only by requiring a K–12 outreach program as a part of their project funding starting in 2005. As the funding parameters only included universities, TNTG goals were revised to meet new challenges and directions. Since the curriculum developed as outreach was government funded, it became a publicly-owned resource. Each university or nano center would post outreach programs on its website, but they were difficult to find, especially since students and teachers were not informed. Thus, TNTG made the decision to gather all those resources, along with global resources, and post them on the group website† as a clearinghouse of teaching and learning tools for teachers and students around the world.

An Expanding Global Perspective

The media tours to Switzerland were to continue over the next 6 years and were comprehensive learning experiences condensed into 8 days each, with overviews published‡ on each trip that covered micro/nanotechnologies, environmental technologies, aerospace and military technologies, and micro manufacturing between 2002 and 2008.

An invitation to Light Feather to be a keynote speaker in Thailand for the inaugural Human Resources Conference concerning nanotechnology, hosted by the Asian Institute of Technology in 2003, prompted the sharing of ideas for research and development courses that would enhance their workforce. The theme of her address was that developing countries must find their current niche in commercial products

---

* UVA Virtual Nanolab: http://www.virlab.virginia.edu/.
† The NanoTechnology Group: http://www.TNTG.org.
‡ http://www.tntg.org/documents/52.html (our work).

and industries as starting points for nanoscience research to improve those areas. In Thailand it was fabrics; therefore, materials science at the nanoscale was the key to building their success.

Among their new products are fabrics and Thai herbal medicines that are both "nanocoated."[*] The Thailand National Nanotechnology Center uses the phrase "to create niche products, and processes with nanotechnology" as their byline.

In 2005, Light Feather attended as the leader of the education focus group designated Working Group 2, at the International Center for Science and High Technology (ICS)—United Nations Industrial Development Organization (UNIDO) North-South Dialogue on Nanotechnology: Challenges and Opportunities[†] in Trieste, Italy. This work led to the publication of the Working Group 2 paper in February 2008:

Title of ICFAI'S Professional Reference Book: *Nanotechnology Issues and Challenges*

The ICFAI (Institute of Chartered Financial Analysts of India) University is a nonprofit organization involved, primarily, in imparting quality education in the area of management and finance. ICFAI Books and Icfai Publications, wings of ICFAI University, are involved in publishing digests, magazines, journals, and executive reference books for professional students, research scholars, academicians, and corporate executives across the world.

The First International Collaboration in the United States on K–12 Nanoscience Courses[‡]

A long-held dream of a team of engineering professors and teachers from Taiwan to visit the United States and share their vision for K–12 nanoscience education was a very informative experience for the participants from both countries. This group had completed animations, videos, comic books, coloring books, interactive video games, and a set of textbooks titled *Nano Symphony for Biology, Chemistry and Physics* for the middle and high school grades. Arizona State University and University of Wisconsin at Madison MRSEC were the hosts for this tour facilitated by the NanoTechnology Group.

---

[*] Article: www.scidev.net/en/new-technologies/thailand-nanotech-plan-moves-ahead.html.

[†] Reports: http://www.tntg.org/documents/52.html.

[‡] Reports: http://www.tntg.org/documents/52.html.

New Book Published in December 2010

The request for Light Feather to author the book *Nanoscience Education, Workforce Training and K–12 Resources* (ISBN: 978-1-4200-5394-4, CRC Press) allowed members of TNTG to work together and suggest a new holistic bottom-up approach for teaching nanoscience in K–12. The coauthor, Miguel F. Aznar, provides the teaching perspective to the book of resources and tools, explaining how to use the ICE-9 method for understanding and evaluating all technologies.

The membership contributed to the workforce training section with overviews of their programs that have initiated the training of nano-technicians for the next wave of nanomanufacturing. NSF will be concentrating its funding for universities to address nanomanufacturing in the 2012 budget with plans for approximately six new specialized centers. The next wave of manufacturing will not be from the industrial era; it will be at the nanoscale of discovery.

Light Feather and TNTG members continue to receive requests to contribute to education publications, such as providing an education chapter for a nanoprocessing book, as a coauthor with Michael Richey at Boeing Corporation and Robert Cormia, one of the contributors for workforce education in her book.

Requests for articles in magazines and help with an education book for India are still in process, and Light Feather and her contributors look forward to reviewing the new standards for science in K–12 developed by experts from the National Academy of Science and Achieve, Inc., in the near future.

Why Do Children Need to Learn Nanoscience?

"This is the question most often asked by grant reviewers and teachers, and in order to answer it we must view all science as an interconnected web of nature that surrounds us. The following answer may give us all a new perspective on teaching science as a holistic ecosystem of nature.

"Science is the study of nature and how the world works. The advances in microscopy over the past two decades have allowed our scientific communities to see into the atomic level, move and manipulate atoms, and create new advances in all branches of science from the microscopic to the cosmos. Nanoscale science is the size where we can see the underlying energy of atoms and particles before they become matter. If we were to teach our young students this size of science, allowing them to compare the visual elements of the atomic scale, in relation to everything they see in the macro scale, it would be taught as 'the foundation of nature'. It is not a separate subject to be added—it is a size—that is of extreme importance in understanding the patterns and relationships of nature that surround us in our everyday lives. Therefore, it is our obligation to introduce this size of science to all students with visual elements that show atoms in movement at the nanoscale as the 'foundation of nature' before it becomes matter."

Conclusions

"Nanoscience has taken the world by storm, and we need to encourage acceptance in our schools if we are ever to compete globally," stated Light Feather. "In conclusion we must create a learning ecosystem that can change our worldview and put the 'wow' back in our classrooms. Life-long learning should be the goal for every human being on the planet, so let us start with our children and gain a holistic view with them, encouraging their innate curiosity and inquiry, inspiring them to share new ideas, challenging them to think and explore nature because they are a part of it."

According to the Webster's dictionary, a visionary deals with utopian ideals and quixotic illusionary concepts, while a realist is defined as enmeshed in the genuine, the authentic, and the factual. If you combined these antithetical viewpoints in one person, you would find Judith Light Feather, a visionary realist. Thus, impelled by the concept of Internet learning and teaching, with passion and perseverance, this practical visionary began the journey of exploring the potential and possibilities of the complex educational system to bring that vision into reality. The story is unusual in that the group has never received any outside funding, but has succeeded in developing a reputation as a respected global voice for nanoscience education solutions.

Judith Light Feather, President
The NanoTechnology Group, Inc.
www.TNTG.org
www.NanoNEWS.TV
Judith.LightFeather@TNTG.org

11

Conclusion: Social Aspects of Nanotechnology

We are living in one of the most exciting times in history, at the edge of the "new technical frontier," watching new worlds created at a microscale that was a sci-fi fantasy just a few short years ago. Our evolution is having a hard time keeping up with the change, as Moore's law is moving technology so fast beyond our personal capacity to mature and understand. Our evolutionary cycle is 20 years, while technology's cycle is 18 months or less. This brings with it a chaotic confusion on one hand, and yet an economic boom on the other.

At no other time in history will the necessity of new job creation be as rapid as we will see in the next few years. In the United States alone, Mihall Roco projected that 0.08–0.09 million nanotechnology workers will be needed by 2015. U.S. policy makers have known of the educational deficit for the past decade. Our population is behind the eight ball in this. We are falling behind as our world speeds up, bringing yet another new job to the table—educating the masses on nanotechnology. We need secretaries, waste management experts, scientists, teachers, policemen, antiterror agents, politicians, policy writers, attorneys, insurance agents, patent people, doctors and nurses, educators, etc.—all with a knowledge of nanotechnology, its power, and its dangers, to work in the new industrial age of nanotechnology (Figure 11.1).

New industries are being formed and traditional enterprises are being visibly changed around the technology. The following are just a few that have already shown the need:

1. Medical and biotechnology
2. Electronics/semiconductor
3. Pharmaceutical
4. Water purification and food industry
5. Optoelectronics
6. MEMS
7. Forensics
8. Education
9. University research

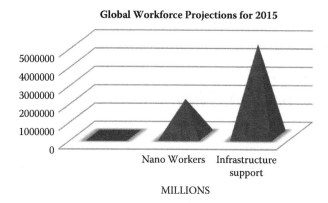

FIGURE 11.1

Two million nanotechnology workers needed worldwide. Five million additional "infrastructure" support jobs needed in the global market by 2015. (From M.C. Roco, Nature Biotechnology, 21(10), 2003.)

A nanotechnology career can include anything from design, fabrication, patent law, research, sales, testing, tech support, PR and marketing, public service, management, distribution, logistics, etc. The possibilities are astounding, and yet a large portion of the population has never heard of nanotechnology in 2011. The 2015 trillion dollar projection is nearly upon us—we have a large learning curve ahead to educate the masses. It is a key challenge to prepare for the education, training, and development of a new generation of skilled workers so that the technology can move forward.

Our educational system will have to advance heavily in the sciences: biology, chemistry, physics, science education, and engineering with all their subcategories, i.e., aerospace, agricultural, biotechnology, chemical, computer, electrical, industrial, nuclear, and materials. Our business departments will have to embrace the management, sales, marketing, logistics, etc., for this grand undertaking. As seen in this text, laws need to change, opening up new fields of law, insurances, and standardization.

We are trying to plug a new expansive technology into the psychosocial conditioning of our lives, and it is a hard fit for many. Nanotechnology is too large, our cultured lives too closed to the idea of tabletop creation. One will have to give.

The increasing demands of accountability and public acceptance of an educated populace are being referred to as new politics. Multiple generations are becoming computer savvy and are prepared to ask the questions on subjects of concern. The populace is informed and involved. This carries directly over to the development of new technologies, specifically nanotechnology.

It is becoming a whole new world, and this author, being a planner and goal setter, am wondering if we moved too fast to set the goals with regard to the new technology. At this point, people are trying to get a handle on it, invest, create new businesses, and enjoy the miracles that nanotechnology innovations are bringing us.

Our lack of standardization in nanotechnology in laws as well as language and definition will create an increasing number of problems for industry, researchers, and policy makers as conflicting definitions and nanotechnology confusion will impact such areas as health and the environment.

As the development of products grows without this international standardization, society will, as it has so many times in the past, serve as a laboratory for the nanotechnology products coming out. Standards in the United States are, in most cases, higher than in some other countries, but we are a global society and consumer driven, grabbing for the newest *bling* without researching the environment in which it is made, supposing that the products meet our U.S. Department of Agriculture (USDA) regulations. Except for a chosen few self-instructed nanotechs, we do not even think of the residual risks.

We need to make sure that weaker societies do not become the laboratory for the rich in the production of the new products. Our standards need to go with the companies to the third world countries that would be manufacturing hubs because of the lack of regulation in those countries.

Nanotechnology scientists and social scientists need to converse to create an understanding and build public trust by educating the populace as a whole. Someone needs to create the plan of implementing this wonderful new technology with all its warts and baubles. Nanotechnology is here to stay. It is our calling as an educated populace to keep it growing and safe, on the path of good for all the individual societies of the world.

Glossary

Active shield: A defensive system with built-in constraints to limit or prevent its offensive use.

Adhesion: Property of certain dissimilar molecules that cling together due to attractive forces.

Aerobots (aerobotics): Aerial (flying) robots.

Atom: Basic unit of matter consisting of a dense, central nucleus surrounded by a cloud of negatively charged electrons. The atomic nucleus contains a mix of positively charged protons and electrically neutral neutrons.

Amino acids: Organic molecules that are the building blocks of proteins. There are some 200 known amino acids, of which 20 are used extensively in living organisms.

Antioxidants: Chemicals that protect against oxidation, which causes rancidity in fats and damage to DNA.

Artificial intelligence (AI): A field of research that aims to understand and build intelligent machines; this term may also refer to an intelligent machine itself.

Assembler: A molecular machine that can be programmed to build virtually any molecular structure or device from simpler chemical building blocks. Analogous to a computer-driven machine shop. *See* Replicator.

Atom: The smallest particle of a chemical element (about three ten-billionths of a meter in diameter). Atoms are the building blocks of molecules and solid objects; they consist of a cloud of electrons surrounding a dense nucleus a hundred thousand times smaller than the atom itself. Nanomachines will work with atoms, not nuclei.

Atomic force microscope (AFM): A very high-resolution microscope that uses a microcantilever to scan the surface of a substrate. This microscope can image and scan surface features on the order of less than a nanometer. Same as scanning force microscope.

Autogenous control: In medical nanorobotics, the conscious control of *in vivo* nanorobotic systems by the human user or patient; in biochemistry, the action of a gene product that either inhibits (negative autogenous control) or activates (positive autogenous control) expression of the gene coding for it.

Automated engineering: The use of computers to perform engineering design, ultimately generating detailed designs from broad specifications with little or no human help. Automated engineering is a specialized form of artificial intelligence.

Bacteria: One-celled living organisms, typically about one micron in diameter. Bacteria are among the oldest, simplest, and smallest types of cells.

Bandgap (in solid-state physics and related fields): The energy range in a solid in which no electron states exist. For insulators and semiconductors, the band gap generally refers to the energy difference (in electron volts) between the top of the valence band and the bottom of the conduction band; it is the amount of energy required to free an outer shell electron from its orbit about the nucleus to a free state.

Baronatation: In medical nanorobotics, locomotion through a frozen fluid by applying mechanical pressure along the path traveled to induce melting ahead, followed by regelation (refreezing) behind.

Biochauvinism: The prejudice that biological systems have an intrinsic superiority that will always give them a monopoly on self-reproduction and intelligence.

BioMEMS: Biological microelectromechanical systems are MEMS systems with applications for the biological/analytical chemistry market.

Biometics: The study relating to the adoption of good designs seen in living beings.

Bionanotechnology: Molecular motors, biomaterials, single-molecule manipulation technologies, biochip technologies, etc.

Biosensor: A device that combines a biological indicator with an electrical, mechanical, or chemical sensing system.

Biostasis: A condition in which an organism's cell and tissue structures are preserved, allowing later restoration by cell repair machines.

Buckyball: Fullerene (a family of molecules composed entirely of carbon, in the form of a hollow sphere, ellipsoid, tube, or plane) forming a spherical shape.

Bulk micromachining: The tailoring of structures by machining a wafer's interior using wet chemical techniques and differential etching rates of different crystallographic planes.

Bulk properties: Material properties that are exhibited when the material is available in large quantities, and largely affected by nanoscale interactions.

Bulk technology: Technology based on the manipulation of atoms and molecules in bulk, rather than individually; most present technology falls in this category.

Capillaries: Microscopic blood vessels that carry oxygenated blood to tissues.

Carbon nanotubes: Tiny tubes about 10,000 times thinner than a human hair and consisting of rolled up sheets of carbon hexagons. A form of carbon with a nanostructure that can have a length:diameter ratio of up to 28,000,000:1. This ratio is significantly larger than in any other material. These cylindrical carbon molecules have novel properties that make them potentially useful in applications in nanotechnology, electronics, optics, and other fields of materials science.

Catalytic converter: Component on the exhaust system of an internal combustion engine used to detoxify harmful emissions before exposing to environment. This device is chemical-reaction driven and catalyzed with a precious metal such as platinum.

Cell: A membrane-bound unit, typically microns in diameter. All plants and animals are made up of one or more cells (trillions, in the case of human beings). In general, each cell of a multicellular organism contains a nucleus holding all of the genetic information of the organism.

Cell engineering: Deliberate artificial modifications to biological cellular systems on a cell-by-cell basis.

Cell repair machine: A system including nanocomputers and molecular scale sensors and tools, programmed to repair damage to cells and tissues.

Cell surgery: In medical nanorobotics, modifying cellular structures using medical nanomachines.

Ceramic: An inorganic, nonmetallic solid prepared by the action of heat and subsequent cooling. A hard porous nonmetallic composite that can exhibit various material properties such as ferroelectricity and superconducting.

Charge-coupled device: A device utilizing a technique in which information is stored and transported by means of packets of minute electrical charges.

Chemical vapor deposition: The growth of thin solid films on a substrate as the result of thermochemical vapor phase reactions. A chemical process used to produce high-purity, high-performance solid materials. The process is often used in the semiconductor industry to produce thin films.

Chemotactic nanosensor: In medical nanorobotics, a nanosensor used to determine the chemical characteristics of surfaces, possibly configured as a pad coated with an array of reversible, perhaps reconfigurable, artificial molecular receptors.

Chip: *See* integrated circuit.

Chirality: A phenomenon is said to be chiral if it is not identical to its mirror image. Here we refer to a molecular direction property that designates a "left hand" and a "right hand" direction where the two symmetries cannot be superposed upon one another.

Chronocyte: In medical nanorobotics, a theorized mobile, mass storage (nanorobotic) device, similar to a communicyte, that may be used as a mobile source of precisely synchronized universal time inside the human body.

Clottocytes: Artificial mechanical platelets (artificial platelets).

Colloid: A chemical mixture where one substance is dispersed evenly within another. The particles of the dispersed substance are only suspended in the mixture, unlike a solution, where they are completely dissolved within.

Communicyte: In medical nanorobotics, a theorized mobile, mass storage (nanorobotic) device that can be used for information transport throughout the human body.

Conductivity: A measure of a material's ability to conduct electric current.

Conjugation: In medical nanorobotics, the docking of two or more nanorobots for the purpose of exchanging information, energy, or materials, or to establish a larger multirobotic structure; in biology, the union of two unicellular organisms accompanied by an interchange of nuclear material, as in *Paramecium*.

Creative destruction: Creative destruction occurs in an industry when a new technology or product paradigm replaces the old. The industry becomes redefined and uses a new technology trajectory. This has occurred in microsystems when MEMS-based accelerometers eclipsed the microelectromechanical systems developed for airbag exploders in passenger cars.

Cross-linking: A process forming chemical bonds between two separate molecular chains.

Cryobiology: The science of biology at low temperatures; research in cryobiology has made possible the freezing and storing of sperm and blood for later use.

Crystal lattice: The regular three-dimensional pattern of atoms in a crystal.

Crystallescence: In medical nanorobotics, the crystallization of solid solute that is offloaded by nanorobot sorting rotors at a concentration that exceeds the solvation capacity of the surrounding solvent.

Crystalline: A solid material whose constituent atoms, molecules, or ions are arranged in an orderly repeating pattern extending in all three spatial dimensions.

Cytocarriage: In medical nanorobotics, the commandeering of a natural motile cell, by a medical nanorobot, for the purposes of *in vivo* transport (of the nanorobot), or to perform a herding function (of the affected cell), or for other purposes.

Cytocide: The killing of living cells.

Cytography: A physical description (and mapping) of the living cell.

Cytoidentification: Identification of cell type.

Cytometrics: The quantitative measurement of cell sizes, shapes, structures, and numbers.

Cytonatation: In medical nanorobotics, swimming around inside a living cell.

Cytonavigation: In medical nanorobotics, navigation inside the cell; cellular navigation.

Cytopenetration: In medical nanorobotics, entry into cells by penetrating the plasma membrane.

Cytoskeletolysis: In medical nanorobotics, purposeful destruction of the cellular cytoskeleton by a nanorobot, for cytocidal purposes.

Cytotomography: Tomographic imaging of an individual cell.

Cytovehicle: In medical nanorobotics, a living cell that has been commandeered by a medical nanorobot for use during cytocarriage.

Deep reactive ion etching: An etching technique that uses plasmas to obtain high-aspect-ratio structures or deep features.

Dendrite (crystal): A crystal that grows in a snowflake pattern or a tree branching pattern.

Deoxyribonucleic acid (DNA): A nucleic acid that contains the genetic instructions used in the development and functioning of all known living organisms and some viruses. DNA molecules are long chains consisting of four kinds of nucleotides; the order of these nucleotides encodes the information needed to construct protein molecules. These in turn make up much of the molecular machinery of the cell. DNA is the genetic material of cells.

Dermal zippers: *See* zippocytes.

Designing ahead: The use of known principles of science and engineering to design systems that can only be built with tools not yet available; this permits faster exploitation of the abilities of new tools.

Design diversity: A form of redundancy in which components of different designs serve the same purpose; this can enable systems to function properly despite design flaws.

Design for manufacturability: Statistical information on manufacturing process characteristics used to ensure that the device design falls within the parameters of normal manufacturing variances for each process element.

Design rules: Rules for design of a device, established by repeated part fabrication, or materials testing, and includes minimum feature widths, minimum feature spacing, feature overlap dimensions, etch release hole spacing, material characteristics, etc.

Diamondoid: Structures that resemble diamonds in a broad sense; strong, stiff structures containing dense, three-dimensional networks of covalent bonds, formed chiefly from first and second row atoms with a valence of three or more. Many of the most useful diamondoid structures will be rich in tetrahedrally coordinated carbon.

Diamondophagy: Eating diamond.

Dielectric: An insulating material in which electrons are bound and unable to freely move within a substrate. A nonconducting substance, i.e., an insulator.

Disassembler: In molecular nanotechnology, a nanomachine or system of nanomachines able to take an object apart while at each step recording the structure and composition of that object at the molecular level.

Discontinuous innovation: Fundamental and far-reaching product changes that require the users or producers to change.

Disequilibration: In medical nanorobotics, maintenance or inducement of a state of perpetual ionic, chemical, or energetic disequilibrium in a living cell by a medical nanorobot, usually for the purpose of inducing cytocide.

Disruptive technology: Those technologies that have redefined the technology/product paradigm in the existing application area and have created the basis for a new industry.

Dissolution: Deterioration in an organism such that its original structure cannot be determined from its current state.

Effervescence: In medical nanorobotics, bubble formation by a gaseous solute that is offloaded by nanorobot sorting rotors at a concentration that exceeds the solvation capacity of the surrounding solvent.

Elasticity: A material property that allows deformation under stress and reformation when stress is released.

Electron: Subatomic particle that carries negative electric charge. The number of electrons in an atom and their energy levels determine many of the electrical properties of material. The electron is not known to have substructure; that is, it is not known to be made up of smaller particles.

Electron beam lithography: The practice of scanning a beam of electrons in a patterned fashion across a surface covered with a film (called the *resist*) (exposing the resist) and of selectively removing either exposed or nonexposed regions of the resist (developing). The purpose is to create very small structures in the resist that can subsequently be transferred into another material for a number of purposes, for example, for the creation of very small electronic devices.

Energy: A scalar physical quantity that describes the amount of work that can be performed by a force. Several different forms of energy exist, including kinetic, potential, thermal, gravitational, sound, light, elastic, and electromagnetic energy.

Engineering: The use of scientific knowledge and trial and error to design systems.

Engulf formation: In medical nanorobotics, a configuration that may be adopted by a metamorphic nanorobot, in which the nanorobot reshapes itself to create an interior cavity capable of trapping a living cell, virion, or other biological particle.

Entropy: A measure of the disorder of a physical system.

Enzyme: A protein that acts as a catalyst in a biochemical reaction.

Eurisko: A computer program developed by Professor Douglas Lenat that is able to apply heuristic rules for performing various tasks, including the invention of new heuristic rules.

Evolution: A process in which a population of self-replicating entities undergoes variation, with successful variants spreading and becoming the basis for further variation.

Exploratory engineering: Design and analysis of systems that are theoretically possible but cannot be built yet, owing to the limitations of currently available tools.

Exponential growth: Growth that proceeds in a manner characterized by periodic doublings.

Fab: The informal name for a chip manufacturer's fabrication plant where ICs or MEMS devices are made. SEMICON industry term for a foundry.

Fact forum: A procedure for seeking facts through a structured, arbitrated debate between experts.

Ferroelectricity: A material property that is characterized by natural electric polarizability that can be altered by an external electric field. The term is used in analogy to *ferromagnetism*, in which a material exhibits a permanent magnetic moment.

Filtration: A mechanical or physical operation that is used for the separation of solids from fluids (liquids or gases) by interposing a medium through which the fluid can pass, but the solids (or at least part of the solids) in the fluid are retained.

Finite element analysis: A simulation procedure for analyzing multiphysics behavior.

Free radical: A molecule containing an unpaired electron, typically highly unstable and reactive. Free radicals can damage the molecular machinery of biological systems, leading to cross-linking and mutation.

Fullerene: A family of molecules composed entirely of carbon, in the form of a hollow sphere, ellipsoid, tube, or plane. Spherical fullerenes are also called *buckyballs*, and cylindrical ones are called carbon nanotubes or *buckytubes*. Graphene is an example of a planar fullerene sheet.

Functional navigation: In medical nanorobotics, a form of nanorobotic navigation in which nanodevices seek to detect subtle variations in their environment, comparing sensor readings with target tissue/cell profiles and then congregating wherever a precisely defined set of preconditions exists.

Fuzzy logic: A method to mathematically represent uncertainty and ambiguity and provide formalized tools to deal with data whose boundaries are not sharply defined (i.e., are fuzzy). Some PalmTops use fuzzy logic to recognize handwriting.

Heisenberg uncertainty principle: A quantum mechanical principle with the consequence that the position and momentum of an object cannot be precisely determined. The Heisenberg principle helps determine the size of electron clouds, and hence the size of atoms.

Heuristics: Rules of thumb used to guide one in the direction of probable solutions to a problem.

High-aspect-ratio micromachining: Micromachining techniques for manufacturing microstructures of aspect ratios.

Histonatation: In medical nanorobotics, locomotion (swimming) through tissues by a nanorobot.

Histonavigation: In medical nanorobotics, navigation through tissues by a nanorobot.

Hydrophilicity/hydrophilic: Hydrophilicity is the tendency of a molecule to be solvated by water. Hydrophile, from the Greek (*hydros*) "water" and φιλια (*philia*) "friendship" refers to a physical property of a molecule that can bond with water.

Hydrophobicity/hydrophobe: The word *hydrophobicity* is obtained from combining the words for a form of water (*hydro*) and for fear (*phobos*) in Attic Greek. It refers to the physical property of a molecule (known as a *hydrophobe*) that is repelled from a mass of water.

Hypertext: A computer-based system for linking text and other information with cross-references, making access fast and criticisms easy to publish and find.

Hypsithermal limit: The maximum amount of energy that may be released at earth's surface, as a result of human technological activities, without significantly altering the natural global energy balance; estimated as 10^{13} to 10^{15} watts.

In cyto: Within a biological cell.

In messaging: In medical nanorobotics, conveyance of information from a source external to the human body, or external to working nanodevices, to a receiver located inside the human body.

Innovation: The exploitation of new ideas to develop novel product processes, services, or business models.

In nucleo: Within the nucleus of a cell.

In sanguo: Within the bloodstream.

Interconnection: A series of connections and interconnections is required in moving from the nano-sized domain to the macro-sized domain where humans communicate with and use products.

Integrated circuit (IC): An electronic circuit consisting of many interconnected devices on one piece of semiconductor, typically into 10 millimeters on a side. ICs are the major building blocks of today's computers.

Integration: Bringing parts together to make a unified whole.

Invention: The creation of a new idea or concept.

Ion: An atom with more or fewer electrons than those needed to cancel the electronic charge of the nucleus. An ion is an atom with a net electric charge.

Kevlar™: A synthetic fiber made by E. I. du Pont de Nemours & Co., Inc. Stronger than most steels, Kevlar is among the strongest commercially available materials and is used in aerospace construction, bulletproof vests, and other applications requiring a high strength-to-weight ratio.

Kinetic energy: The kinetic energy of an object is the extra energy that it possesses due to its motion. It is defined as the work needed to accelerate a body of a given mass from rest to its current velocity.

Laser: Abbreviation of light amplification of stimulated emission radiation. A device that emits light (electromagnetic radiation) through a process called *stimulated emission*. Laser light is usually spatially coherent, which means that the light either is emitted in a narrow, low-divergence beam, or can be converted into one with the help of optical components such as lenses.

Lightsail: A spacecraft propulsion system that gains thrust from the pressure of light striking a thin metal film.

Limited assembler: An assembler with built-in limits that constrain its use (for example, to make hazardous uses difficult or impossible, or to build just one thing).

Liquid crystal display (LCD): An electronically modulated optical device shaped into a thin, flat panel made up of color or monochrome pixels filled with liquid crystals. It is arrayed in front of a light source (backlight) or reflector and is often used in battery-powered electronic devices because it uses very small amounts of electric power.

Lithography (photolithography): Method of fabrication of integrated circuits and microelectromechanical systems that uses alternating steps of material deposition and removal. The process selectively removes parts of a thin film or the bulk of a substrate. It uses light to transfer a geometric pattern from a photomask to a light-sensitive chemical photoresist on the substrate. A series of chemical treatments then engrave the exposure pattern into the material underneath the photoresist.

Macrosensing: In medical nanorobotics, the detection of global somatic states (inside the human body) and extrasomatic states (sensory data originating outside of the human body) by *in vivo* nanorobots.

Magnetic memory: Storage of information on magnetized material such as magnetic tape. Magnetic storage uses different patterns of magnetization in a magnetizable material to store data and is a form of nonvolatile memory (a memory that will retain the stored information even if it is not constantly supplied with electric power). An example of a magnetic memory is a computer hard disk drive.

Massometer: In medical nanorobotics, a nanosensor device for measuring the mass of individual molecules or small physical objects to single-proton resolution.

Materials science: An interdisciplinary field involving the properties of matter and its applications to several areas of science and engineering. Materials science investigates the relationship between the structure of materials at atomic or molecular scales and their macroscopic properties. It includes elements of applied physics and chemistry, as well as chemical, mechanical, civil, and electrical engineering.

Melting point: The temperature point in which a material transitions from solid to liquid.

MEME: An idea that can replicate and evolve like a gene. Examples of MEMEs (and MEME systems) include political theories, proselytizing religions, and the idea of MEMEs themselves.

Messenger molecule: A chemically recognizable molecule that can convey information after it is received and decoded by an appropriate chemical sensor.

Metamorphic: In medical nanorobotics, capable of adopting multiple physical configurations via smooth changes from one configuration to another.

Microbiotagraphics: Mapping the microbiotic populations present in the human body.

Microbivores: Artificial mechanical phagocytes (artificial white cells).

Microelectromechanical system (MEMS): Term primarily used in the United States for a micro-sized device with both electrical and mechanical functionality.

Microelectronics: This technology includes techniques used to manufacture ICs, discrete microelectronic devices, MEMS devices such as sensors and actuators, and various electro-optic devices.

Microelectro-optical mechanical systems: MEMS devices that have applications in optical telecommunications. Also known as optical MEMS.

Microfabrication: A manufacturing technology for making microscopic devices, such as integrated circuits or MEMS.

Microfluidics: The science of designing, manufacturing, and formulating devices and processes that deal with the nanoliter or picoliter volumes of fluids (i.e., 10^{-9} or 10^{-12} liters.) Microfluidic studies include nozzles, pumps, reservoirs, mixers, valves, etc., that can be used for a variety of applications, including drug dispensing, ink-jet printing, and general transport of liquids, gases, and their mixtures.

Micromachines: Mechanical devices with microscopic dimensions. This term is preferred in Japan and is used interchangeably with MEMS and MST.

Micromanufacturing: The production of microsystems products using microfabrication, packaging, microassembly, and other technologies associated with microelectronics and microsystems.

Micropackaging: The processes used to make the connections/interconnections, encapsulate, and protect the MEMS/microdevice or nanodevice or subsystem so that it is ready to be microassembled into a product usable in the macroworld.

Microsystem: A microscale device that combines some of the following functions: mechanical, optical, chemical, thermal, magnetic, biological, and fluidic, generally integrated with electronics. A microsystem is a packaged assembly where all the connections are in place to interface with the macroworld and is usable by the consumer.

Microsystem technology (MST): The term *MST* is preferred in Europe and Japan in place of MEMS as it is the methods and techniques used to fabricate microsystems.

Microtechnology: Technology dealing with matter on the size scale of microns (1 millionth of a meter). Microtechnology is a broad term and can refer to microelectronics, MEMS, or any technology that manipulates matter on a micron scale.

Millipede memory: A nonvolatile memory developed by IBM that uses nanoimprints to code information and atomic force sensing to decode information. It promises a data density of more than 1 terabit per square inch (1 gigabit per square millimeter), about four times the density of magnetic storage available today.

Molecular assembler: A general purpose device for molecular manufacturing, able to guide chemical reactions by positioning individual molecules to atomic accuracy (e.g., mechanosynthesis) and to construct a wide range of useful and stable molecular structures according to precise specifications.

Molecular manufacturing: Manufacturing using molecular machinery, giving molecule-by-molecule control of products via positional chemical synthesis, to produce complex molecular structures manufactured to precise specifications.

Molecular medicine: A variety of pharmaceutical techniques and gene therapies that address specific molecular diseases or molecular defects in biological systems.

Molecular nanoscience: An emerging interdisciplinary field that combines the study of molecular/biomolecular systems with the science and technology of nanoscale structures.

Molecular nanotechnology: Thorough, inexpensive control of the structure of matter based on molecule-by-molecule control of products and by-products; the products and processes of molecular manufacturing, including molecular machinery; a technology based on the ability to build structures to complex, atomic specifications by mechanosynthesis or other means; most broadly, the engineering of all complex mechanical systems constructed from the molecular level.

Molecular sorting rotor: A class of nanomechanical device capable of selectively binding (or releasing) molecules from (or to) solution, and of transporting these bound molecules against significant concentration gradients.

Molecular surgery (molecular repair): In medical nanorobotics, the analysis and physical correction of molecular structures in the body using medical nanomachines.

Molecular technology: *See* nanotechnology.

Molecule: A unit of two or more atoms held together by covalent bonds (a form of chemical bonding that is characterized by the sharing of pairs of electrons between atoms, or between atoms and other covalent bonds).

Monkeywrenching: In medical nanorobotics, the mechanical or chemical jamming of cellular equilibrium processes, with a cytocidal objective. *See also* disequilibration.

Moore's law: The number of transistors the industry would be able to place on an IC chip would double every 18 months. The IC industry has been able to keep pace with this law to date and strives to achieve the same in the future. Named after cofounder of Intel, Gordon Moore.

Mutation: An inheritable modification in a genetic molecule, such as DNA. Mutations may be good, bad, or neutral in their effects on an organism; competition weeds out the bad, leaving the good and the neutral.

Nanapheresis: In medical nanorobotics, the removal of blood-borne medical nanorobots from the body using aphersis-like processes.

Nano: A prefix meaning 10 to the minus 9th power, or 1 billionth.

Nanobiosensor: A device that combines a biological indicator with an electrical, mechanical, or chemical sensing system on the nanoscale.

Nanocentrifuge: In medical nanorobotics, a proposed nanodevice that can spin materials at very high speed, imparting rotational accelerations of up to 1 trillion gravities (g's), thus permitting rapid sortation.

Nanochronometer: In medical nanorobotics, a proposed clock or timing mechanism constructed of nanoscale components.

Nanocomputer: A computer made from components (mechanical, electronic, or otherwise) on a nanometer scale.

Nanocrit (Nct): In medical nanorobotics, volume fraction or bloodstream concentration of medical nanorobots, expressed as a percentage.

Nanocrystal: A single crystalline material that has one dimension on the order of 100 nm.

Nanomachines: Biomacromolecules, which are nanoengines acting both as thermal engines and as informational engines, like the so-called assemblers (cf. molecular nanotechnology). The latter are not self-programmed like nanobiomachines.

Nanomanufacturing: The production of nanosystems products using nanofabrication, packaging, nanoassembly, and other technologies associated with nanoelectronics and nanosystems.

Nanomaterial: A material that exhibits distinct properties when studied on the order of less than 100 nm.

Nanomedicine: (1) The comprehensive monitoring, control, construction, repair, defense, and improvement of all human biological systems, working from the molecular level, using engineered nanodevices and nanostructures. (2) The science and technology of diagnosing, treating, and preventing disease and traumatic injury, of relieving pain, and of preserving and improving human health, using molecular tools and molecular knowledge of the human body. (3) The employment of molecular machine systems to address medical problems, using molecular knowledge to maintain and improve human health at the molecular scale.

Nanometer (nm): A unit of measure equal to 1 billionth of a meter.

Nanoparticles: Both synthetic (bottom-up) and transformative (top-down) fabrication rely on the availability of building block materials and artifacts such as quantum dots, nanotubes and nanofibers, ultrathin films, and nanocrystals. Particles with size on the order of 1–100 nm.

Nanopore: A small pore in an electrically insulating membrane that can be used as a single-molecule detector.

Nanoscale: A term used to refer to objects with dimensions on the order of 1–100 nm.

Nanorobot: A computer-controlled robotic device constructed of nanometer-scale components to molecular precision, usually microscopic in size (often abbreviated as nanobot).

Nanosensor: A chemical or physical sensor constructed using nanoscale components, usually microscopic or submicroscopic in size.

Nanosieving: In medical nanorobotics, a nanodevice that can sort molecules or other nanoscale objects by physical sieving.

Nanosystem: A nanoscale device, constructed atom by atom, that combines or stimulates some of the following functions: mechanical, optical, chemical, thermal, magnetic, fluidic, or electronics.

Nanotechnology: Technology dealing with matter on a molecular size scale, on the order of nanometers (1 billionth of a meter). Technology based on the manipulation of individual atoms and molecules to build structures to complex, atomic specifications.

Nanowhisker: A nanoscale structure that consists of brushes attached along a common spine.

Nanowire: A wire of diameter on the order of a nanometer.

Naturophilia: An exclusive love of nature, disdaining everything that is artificial or technological.

Navicyte: In medical nanorobotics, a mobile, mass storage (nanorobotic) device, similar to a communicyte, that may be used to establish a navigational network inside the human body.

Neural stimulation: Imitating the functions of a neural system, such as the brain, by simulating the function of each cell.

Neuron: A nerve cell, such as those found in the brain.

Nonvolatile memory: A memory that will retain the stored information even if it is not constantly supplied with electric power.

Nucleation: A site in which a phase transition begins and grows outward from. Some examples of phases that may form via nucleation in liquids are gaseous bubbles, crystals, or glassy regions.

Nucleotide: A small molecule composed of three parts: a nitrogen base (a purine or pyrimidine), a sugar (ribose or deoxyribose), and phosphate. Nucleotides serve as the building blocks of nucleic acids (DNA and RNA).

Nucleus: In biology, a structure in advanced cells that contains the chromosomes and apparatus to transcribe DNA into RNA. In physics, the small, dense core of an atom.

Optics: The study of the behavior and properties of light, including its interactions with matter and its detection by instruments.

Optoelectronics: The study and application of electronic devices that source, detect, and control light (including invisible forms of radiation such as gamma rays, x-rays, ultraviolet, and infrared, in addition to visible light). Optoelectronic devices are electrical-to-optical or optical-to-electrical transducers, or instruments that use such devices in their operation.

Organic molecule: A molecule containing carbon; the complex molecules in living systems are all organic molecules in this sense.

Outmessaging: In medical nanorobotics, conveyance of information from a transmitter located inside the human body, especially from working nanodevices, to the patient or to a recipient external to the human body.

Pharmacyte: In medical nanorobotics, a theorized (nanorobotic) device capable of delivering precise doses of biologically active chemicals to individually addressed human body tissue cells (e.g., cell-by-cell drug delivery).

Polarization: A property of waves that describes the orientation of their oscillations.

Polydimethylsiloxane (PDMS): An inorganic polymer used in nanotechnology applications such as nanoimprint and soft lithography. It is the most widely used silicon-based organic polymer.

Polymer: A large molecule (macromolecule) composed of repeating structural units typically connected by covalent chemical bonds. While *polymer* in popular usage suggests plastic, the term actually refers to a large class of natural and synthetic materials with a variety of properties.

Polymer clay: Deformable composite of polyvinyl chloride (PVC) that can be manipulated similar to a clay. This usually does not contain any actual clay.

Positional navigation: In medical nanorobotics, a form of nanorobotic navigation in which nanodevices know their exact location inside the human body to approximate micron accuracy continuously at all times.

Positive sum: A term used to describe a situation in which one or more entities can gain without other entities suffering an equal loss, for example, a growing economy. *See* zero sum.

Presentation semaphore: In medical nanorobotics, a mechanical device used to display specific antigens, chemical ligands, or other molecular objects to the external environment, with the purpose of selectively modifying the chemical or other surface characteristics of a nanorobot exterior.

Quantum dot: A semiconductor in which the electron propagation is confined in three dimensions (differing from *quantum wires*, in which propagation is controlled in two dimensions, and *quantum wells*, in which propagation is controlled in a single direction).

Reactive ion etching: Dry etching by plasma having chemically active gas ions.

Redundancy: The use of more components than are needed to perform a function; this can enable a system to operate properly despite failed components.

Replicator: Any system that can build copies of itself when provided with the appropriate raw materials and energy.

Respirocrit: In medical nanorobotics, the volume fraction or bloodstream concentration of respirocyte nanorobots, expressed as a percentage.

Respirocyte: In medical nanorobotics, a theorized blood-borne spherical 1-micron (nanorobotic) device having a 1,000 atm pressure vessel with active pumping powered by endogenous serum glucose, that serves as a mechanical artificial red blood cell.

Restrictive enzyme: An enzyme that cuts DNA at a specific site, allowing biologists to insert or delete genetic material.

Ribonuclease: An enzyme that cuts RNA molecules into smaller pieces.

Ribonucleic acid (RNA): A molecule similar to DNA. In cells, the information in DNA is transcribed to RNA, which in turn is "read" to direct protein construction. Some viruses use RNA as their genetic material.

Ribosome: A molecular machine, found in all cells, which builds protein molecules according to instructions read from RNA molecules. Ribosomes are complex structures built of protein and RNA molecules.

Sanguination: In medical nanorobotics, locomotion (especially swimming by a nanorobot) through the bloodstream.

Sapphirophagy: Eating sapphire (corundum).

Scanning force microscope: A very high-resolution microscope that uses a microcantilever to scan the surface of a substrate. This microscope can image and scan surface features on the order of less than a nanometer. Same as atomic force microscope.

Scanning tunneling microscope: A widely used instrument for viewing surfaces at the atomic level. It provides a three-dimensional profile of viewed surfaces, which is very useful for characterizing surface roughness, observing surface defects, and determining the size and conformation of molecules and aggregates on the surface.

Science: The process of developing a systematized knowledge of the world through the variation and testing of hypotheses. *See* engineering.

Science court: A name (originally applied by the media) for a government-conducted fact forum.

Sealed assembler laboratory: A work space, containing assemblers, encapsulated in a way that allows information to flow in and out but does not allow the escape of assemblers or their products.

Self-assembly: Processes in which a disordered system of preexisting components forms an organized structure or pattern as a consequence of specific, local interactions among the components, without external direction.

Semiconductor: A material that exhibits electrical conductivity properties between those of a *conductor* and an *insulator*. (Insulators do not conduct well, while metals readily conduct.) The conductivity of a semiconductor material can be varied under an external electrical field. Devices made from semiconductor materials are the foundation of modern electronics, including radio, computers, telephones, and many other devices. Semiconductor devices include the transistor, many kinds of diodes including the light-emitting diode, the silicon-controlled rectifier, and digital and analog integrated circuits. Solar photovoltaic panels are large semiconductor devices that directly convert light energy into electrical energy.

Smart sensor: The electronics associated with a sensor that processes the output and is partially or completely integrated on a single chip.

Solar cell: Semiconductor or organic device used to harness solar energy and convert it into electrical energy.

Solution: A homogeneous mixture composed of two or more substances. In such a mixture, a solute is dissolved in another substance, known as a solvent.

Surface area: A measure of exposed area of an object.

Surface-to-volume ratio: The ratio of exposed surface area to volume of the particle. In nanotechnology, a very high surface-to-volume ratio is the enabler for many nanoscale properties.

Suspension: A heterogeneous fluid containing solid particles that are sufficiently large for sedimentation. An example of a suspension would be sand in water. The suspended particles are visible under a microscope and will settle over time if left undisturbed. This distinguishes a suspension from a colloid, in which the suspended particles are smaller and do not settle.

Synapse: A structure that transmits signals from a neuron to an adjacent neuron (or other cell).

Thermogenic limit: In medical nanorobotics, the maximum amount of waste heat that may safely be released by a population of *in vivo* medical nanorobots that are operating within a given tissue volume.

Top-down nanosystems: These systems utilize metaphors such as "chip" (large-scale integrated nanosystems) or "cell" (small-complex nanosystems that function in larger quantities) that are revolutionary in nature and require a much longer term to develop. They utilize MEMS-based manufacturing technology techniques.

Top-down nanotechnology: Engineers taking existing devices, such as transistors, and making them smaller using microtechnology techniques.

Transtegumental: Crossing or passing through the skin or covering of a body.

Ultraviolet light: Electromagnetic radiation with a wavelength shorter than that of visible light, but longer than x-rays, in the range 10 to 400 nm, and energies from 3 to 124 eV. It is so named because the spectrum consists of electromagnetic waves with frequencies higher than those that humans identify as the color violet.

Vasculocyte: In medical nanorobotics, a theorized (nanorobotic) device capable of performing repairs of an injured vascular luminal surface.

Vasculoid: A single, complex, multisegmented nanotechnological medical robotic system capable of duplicating all essential thermal and bio-chemical transport functions of the blood, including circulation of respiratory gases, glucose, hormones, cytokines, waste products, and cellular components. See *Vasculoid: A Personal Nanomedical Appliance to Replace Human Blood*, by Robert A. Freitas Jr. and Christopher J. Phoenix, April 2002.

Vasculography: A physical description (and mapping) of the human vascular system.

Virus: A small replicator consisting of a little package of DNA or RNA, which, when injected into a host cell, can direct the cell's molecular machinery to make more viruses.

Vitamins (engineering): In machine replication theory, vitamin parts are components of a self-replicating machine that the machine is incapable of producing itself; therefore these vital parts must be supplied from an external source.

Volitional normative model of disease: In medical nanorobotics, disease is said to be present in a human being upon either (1) the failure of optimal physical (e.g., biological) functioning or (2) the failure of desired (by the patient) functioning.

Zero sum: A term used to describe a situation in which one entity can gain only if other entities suffer an equal loss, for example, a private poker game. *See* positive sum.

Zippocytes: In medical nanorobotics, a theorized medical nanorobot that can rapidly perform incision wound repairs to the dermis and epidermis; dermal zippers.

Acronyms

CBI	Confidential business information
CNT	Carbon nanotube
CPSC	Consumer Products Safety Commission
CR-BSI	Catheter-related blood stream infection
DOD	Department of Defense
DOE	Department of Energy
DRAM	Dynamic random access memory. A memory that stores each bit of data in a separate capacitor within an integrated circuit. Since real capacitors leak charge, the information eventually fades unless the capacitor charge is refreshed periodically. Because of this refresh requirement, it is a *dynamic* memory as opposed to SRAM and other types of *static* memory.
EHS	Environment, health, and safety
EU27	Twenty-seven nations of the European Union
FDA	Food and Drug Administration
FET	Field effect transistor
FRAM	Ferroelectric random access memory. A memory that uses a ferro-electric layer rather than a dielectric layer to achieve nonvolatility. A nonvolatile memory will retain the stored information even if it is not constantly supplied with electric power.
GAO	Government Accountability Office
GDP	Gross domestic product
GNP	Gross national product
ICON	International Council on Nanotechnology
IND	Investigational New Drug Application
LAD	Luer-activated device
M&A	Mergers and acquisitions
MRSEC	Materials Research Science and Engineering Centers
NCI	National Cancer Institute
NEHI	Nanotechnology, Environmental, and Health Implications
NIEHS	National Institute of Environmental Health Sciences
NIH	National Institutes of Health
NIOSH	National Institute for Occupational Health and Safety
NISE	Nanoscale Informal Science Education
NIST	National Institute of Standards and Technology
NNAP	National Nanotechnology Advisory Panel
NNCO	National Nanotechnology Coordination Office
NNI	National Nanotechnology Initiative
NRC	National Research Council
NRI	Nanoelectronics Research Initiative

NSEC	National Science and Engineering Center
NSET	Nanoscale Science, Engineering, and Technology Subcommittee of the NSTC
NSF	National Science Foundation
NSTC	National Science and Technology Council
OECD	Organization for Economic Cooperation and Development
OMB	Office of Management and Budget
OSTP	Office of Science and Technology Policy
PCA	Program component area
PCAST	President's Council of Advisors on Science and Technology
R&D	Research and development
SBIR	Small Business Innovation Research
SOT	Statement of task
STTR	Small Business Technology Transfer
SWCNT	Single-walled carbon nanotube
TIP	Technology Innovation Program
VC	Venture capital

Nanotechnology Information Sources

Government

Argonne National Laboratory
  http://www.anl.gov/

Defense Advanced Research Projects Agency (DARPA)
  http://www.darpa.mil/

Department of Defense Nanoscience and Technology
  http://www.nanosra.nrl.navy.mil/
  http://www.nanosra.nrl.navy.mil/muri.php

Department of Energy, Basic Energy Sciences
  http://www.er.doe.gov/bes/suf/user_facilities.html
  http://www.ornl.gov/

Department of Industry, Science and Resources, Australia
  http://www.innovation.gov.au/Pages/default.aspx

Institute for Soldier Nanotechnologies
  http://web.mit.edu/isn/

Lawrence Berkeley National Lab
  http://ncem.lbl.gov/

Los Alamos Neutron Science Center
  http://lansce.lanl.gov/

Nanoscience and Nanotechnology—Joint Study: Royal Society and Royal
  Academy of Engineering, UK
  http://www.nanotec.org.uk/index.htm

Nanotechnology in Australia
  http://www.innovation.gov.au/INDUSTRY/NANOTECHNOLOGY/
    Pages/default.aspx

National Aeronautics and Space Administration Ames Center for
  Nanotechnology

http://www.ipt.arc.nasa.gov/index.html
http://mmptdpublic.jsc.nasa.gov/jscnano/
http://www.nas.nasa.gov/Main/Redirector/redirector.php?old_
    url=http://www.nas.nasa.gov/Groups/SDM/
http://ipt.arc.nasa.gov/nanotechnology.html

National Institute of Standards and Technology (NIST)
http://www.nist.gov/pml/index.cfm
http://www.ewalab.com/projects/inst/nistaml.htm

National Nanotechnology Initiative
http://www.nano.gov/

National Research Council of Canada
http://www.nrc-cnrc.gc.ca/eng/research/nanotechnology.html

Naval Research Laboratory
http://www.nrl.navy.mil/pao/pressRelease.php?Y=2003&R=63-03r

Naval Research Tribology Section
http://www.nrl.navy.mil/chemistry/6170/6176/index.php

Pacific Northwest National Laboratory (PNNL)
http://www.pnl.gov/nano/

Sandia
http://www.sandia.gov/mission/ste/capabilities/cint.pdf

Federal Agencies

Federal Agencies with Budgets Dedicated to Nanotechnology Research and Development

Consumer Product Safety Commission (CPSC)
Department of Defense (DOD)
Department of Energy (DOE)
Department of Homeland Security (DHS)
Department of Justice (DOJ)
Department of Transportation (DOT), including the Federal Highway
    Administration (FHWA)
Environmental Protection Agency (EPA)
Food and Drug Administration (FDA) (Department of Health and Human
    Services)
Forest Service (FS), Department of Agriculture
National Aeronautics and Space Administration (NASA)

National Institute for Occupational Safety and Health (NIOSH),
  Department of Health and Human Services/Centers for Disease Control
  and Prevention
National Institute of Food and Agriculture (NIFA), Department of
  Agriculture
National Institutes of Health (NIH), Department of Health and Human
  Services
National Institute of Standards and Technology (NIST), Department of
  Commerce
National Science Foundation (NSF)

Other Participating Agencies

Bureau of Industry and Security (BIS, Department of Commerce)
Department of Education (DOEd)
Department of Labor (DOL)
Department of State (DOS)
Department of the Treasury (DOTreas)
Director of National Intelligence (DNI)
International Trade Commission (ITC)
Nuclear Regulatory Commission (NRC)
U.S. Geological Survey (USGS, Department of the Interior)
U.S. Patent and Trademark Office (USPTO, Department of Commerce)

Grants/Funding Opportunities

Ames Research Center—unsolicited proposals
  http://server-mpo.arc.nasa.gov/Services/Grants/Includes/Unsolicited.tml

Army Research Laboratory Postdoctorate Fellowship Program through Oak
  Ridge Associated Universities (ORAU)
  http://see.orau.org/ProgramDescription.aspx?Program=10186

Bioengineeing, and Chemical Engineering—NSF/Directorate for
  Engineering
  http://www.nsf.gov/pubs/2003/nsf03568/nsf03568.html

Classified—Radar Absorbent Material Improvement Services and
  Production—Department of the Navy—RFP
  http://www.fbo.gov/spg/DON/NAVSEA/N63394/N6339407R1278/
    listing.html

Cooperative Research Centers—Center Proposals—NSF/Directorate for
  Engineering
  http://www.nsf.gov/pubs/2001/nsf01116/nsf01116.html

Defense Sciences Research and Technology—Defense Advanced Research
  Projects Agency
  http://www.darpa.mil/dso/solicitations/solicit.htm

Department of Defense solicitations open for proposal submission
  http://www.dodsbir.net/solicitation/
DoD solicitation open for proposal submission http://www.
  educationmoney.com
  http://www.dtra.mil
  http://www.federalgrantswire.com/federal-grants-wire-site-directory.html
  http://www.defense.gov/search/

Electromagnetic Protection Materials—RFP
  https://www.fbo.gov/index?s=opportunity&mode=form&id=22652b7a17
    00d2383608b6971ab9f92c&tab=core&_cview=0&cck=1&au=&ck=

Electronics Research Program—Office of Naval Research
  http://www.onr.navy.mil/sci_tech/information/312_electronics/

Electronics, Sensors, and Network Research Program—Office of Naval
  Research
  http://www.onr.navy.mil/en/Science-Technology/Departments/
    Code-31/All-Programs/312-Electronics-Sensors.aspx

Grant Opportunities for Academic Liaison with Industry (GOALI) (seeks to
  fund research that lies beyond that which industry would normally fund
  by themselves)
  http://www.nsf.gov/pubs/2007/nsf07522/nsf07522.htm
    Goddard Space Flight Center—unsolicited proposals
  http://code210.gsfc.nasa.gov/grants/grants.htm

Headquarters—unsolicited proposals
  http://prod.nais.nasa.gov/pub/pub_library/unSol-Prop.html

John C. Stennis Space Center—unsolicited proposals
  http://code210.gsfc.nasa.gov/grants/grants.htm

Kennedy Space Center—unsolicited proposals
  http://prod.nais.nasa.gov/pub/pub_library/unSol-Prop.html

Langley Research Center—unsolicited proposals
  http://prod.nais.nasa.gov/pub/pub_library/unSol-Prop.html

Materials Science Program—Department of the Army
  https://researchfunding.duke.edu/detail.asp?OppID=4616

Materials Science and Technology Research Program—Office of Naval
Research
http://www.onr.navy.mil/sci_tech/engineering/332_materials/

Measurement, Science, and Engineering Research Grants Programs—
Center for Nanoscale Science and Technology (CNST)—National
Institute of Standards and Technology/Technology Administration/
DOC
http://www.nist.gov/cnst/upload/CNST_brochure.pdf

Measurement, Science, and Engineering Research Grants Programs—
Materials Science and Engineering Laboratory (MSEL) Grants
Program—National Institute of Standards and Technology/Technology
Administration/DOC
http://www.nist.gov/msel/working-with-msel.cfm

Mechanical Sciences Program—Department of the Army
http://www.grants.gov/search/search.do?oppId=11441&mode=VIEW

Microsystems Technology—Defense Advanced Research Projects Agency
http://www.darpa.mil/mto/

Nanostructured Dental Composite Restorative Materials (R21)—National
Institute of Dental and Craniofacial Research/NIH/DHHS
http://grants2.nih.gov/grants/guide/rfa-files/RFA-DE-07-005.html

Nano-Technology for Coatings in Coal-Fired Environments—Department
of Energy
http://www.fedgrants.gov/applicants/DOE/PAM/HQ/
DE-PS26-05NT42244-02/listing.html

National Center for Electron Microscopy (NCEM)
http://ncem.lbl.gov/

National Institute of Health
http://grants.nih.gov/grants/guide/pa-files/PA-08-053.html

National Institute of Health—Bioengineering Nanotechnology Initiative
http://grants.nih.gov/grants/guide/pa-files/PA-10-149.html
http://grants.nih.gov/grants/guide/pa-files/PA-10-150.html

National Institute of Health—Nanoscience and Nanotechnology in Biology
and Medicine
http://www07.grants.gov/search/search.do;jsessionid=vDlyJhhCmrnnf7
MSSm6dPQG5TqjbJ0b3y8byMhF5nKMpjtBD0mJF!-108218008?oppId=
44022&flag2006=false&mode=VIEW

National Science Foundation (NSF) proposal submission
  http://www.nsf.gov/funding/
  http://www.nsf.gov/funding/azindex.jsp

NCI Alliance for Nanotechnology in Cancer—NSF, NIH
  http://www.nano.cancer.gov

NSF—Electronics, Photonics and Magnetic Devices
  http://www07.grants.gov/search/search.do;jsessionid=Gx50Mmzf7CWJl
    66KlFGjTHrZLCHxmlPgJLdHjb2d27JC6gndQ9Gh!1347027669?oppId
    =46283&mode=VIEW

NSF—Scalable Nanomanufacturing
  http://www07.grants.gov/search/search.do;jsessionid=Gx50Mmzf7CWJl
    66KlFGjTHrZLCHxmlPgJLdHjb2d27JC6gndQ9Gh!1347027669?oppId
    =58167&mode=VIEW

Partnerships for Innovation Program (PFI)
  http://www.grants.gov/search/search.do;jsessionid=Hy7ZMBTW90mJqb
    0LK5RBKWHvCTpcpxlpSLjDPg4Q1Gl62vNQPdp7!695228645?oppId=
    55577&mode=VIEW

Research Instrumentation Grants—Department of the Army
  http://www.arl.army.mil/www/default.cfm?Action=6&Page=8

Research Interests of the Air Force Office of Scientific Research—Aerospace
  and Materials Sciences—Air Force Office of Scientific Research
  http://www.wpafb.af.mil/library/factsheets/factsheet.asp?id=9196

Research Participation at the U.S. Army Research Laboratory Oak Ridge
  Institute for Science and Education
  http://see.orau.org/ProgramDescription.aspx?Program=10084

Strategic Technologies—Defense Advanced Research Projects Agency
  http://www.darpa.mil/sto/solicitations/index.html

Type AC Grants (PRF)—American Chemical Society
  http://www.chemistry.org/portal/a/c/s/1/acsdisplay.html?DOC=prf\
    prfgrant.html#typeac

Type B Grants (PRF)—American Chemical Society
  http://www.chemistry.org/portal/a/c/s/1/acsdisplay.html?DOC=prf\
    index.html

Type G Starter Grants (PRF)—American Chemical Society
http://www.chemistry.org/portal/a/c/s/1/acsdisplay.html?DOC=prf\
index.html

Undersea Weapons and Naval Materials Research Division—Office of
Naval Research
http://www.onr.navy.mil/en/Science-Technology/Departments/
Code-33/All-Programs/332-naval-materials.aspx

U.S. Department of Health and Human Services
http://grants1.nih.gov/grants/oer.htm
http://grants1.nih.gov/grants/guide/search_results.htm?text_
curr=Nanotechnology&Search_Guide.x=0&Search_Guide.
y=0&scope=pa-rfa&year=active&sort=

Weapons and Materials Research—Department of the Army
http://www.arl.army.mil/www/default.cfm?Action=35&Page=35

Young Investigator Program (YIP)—Department of the Army
http://www.arl.army.mil/main/main/DownloadedInternetPages/
CurrentPages/DoingBusinesswithARL/research/AROBAA1102.doc

Young Investigator Program—Office of Naval Research
http://www.fedgrants.gov/EPSData/USN/Synopses/4/
BAA%26%23032%3B05-002/FY%26%23032%3B2005%26%23032%3B
YIP.doc

Academic Labs and Research Centers

Aarhus University (Denmark)
http://inano.au.dk/

Arizona State University
http://www.eas.asu.edu/%7Emae/
http://www.biodesign.asu.edu/centers/anb/

Australian National University (Australia)
http://wwwrsphysse.anu.edu.au/eme/home.php

Beckman Institute
http://www.beckman.illinois.edu/index.aspx

Boston University
  http://www.bu.edu/dbin/bme/

Brown University
  http://www.cs.brown.edu/people/jes/nano.html

California Institute of Technology (Caltech)
  http://www.wag.caltech.edu/
  http://www.cmp.caltech.edu/~roukes/

Carnegie Mellon University
  http://www.me.cmu.edu/faculty1/higgs/

Case Western Reserve University
  http://dmseg5.mse.cwru.edu/Groups/Ernst/

Center for Economic Growth (New York)
  http://www.nylovesnano.com/assetMap.php?id-113&type-R&D

City University of Hong Kong (Hong Kong)
  http://www.cityu.edu.hk/cityu/dpt-acad/fse-me.htm

Clemson University
  http://www.ch2m.com/corporate/markets/electronics_and_advanced_
    technology/assets/ProjectPortfolio/Clemson.pdf

Columbia University
  http://www.cise.columbia.edu/nsec/research/

Cordis (European Research)
  http://cordis.europa.eu/nanotechnology/

Cornell University
  http://www.mse.cornell.edu/%20
  http://www.cnf.cornell.edu/
  http://www.cns.cornell.edu/
  http://www.nbtc.cornell.edu/

Cranfield University (UK)
  http://www.cranfield.ac.uk/sims/
  http://www.cranfield.ac.uk/sas/nanotech/

Delft University of Technology (The Netherlands)
  http://www.tudelft.nl/live/pagina.jsp?id=2d1e7854-5d37-4e6a-a55f-
    589383446de3&lang=en

Doshisha University (Japan)
  http://engineering.doshisha.ac.jp/english/kenkyu/gakka/d_mse.html

Duke University Ceint
  http://www.ceint.duke.edu/

Eindhoven University of Technology (The Netherlands)
  http://www.schubert-group.com
  http://www.chem.tue.nl

Florida International University
  http://web.eng.fiu.edu/~agarwala

Georgia Institute of Technology
  http://grover.mirc.gatech.edu/
  http://www.me.gatech.edu/
  http://www.nanoscience.gatech.edu/zlwang/

Harvard University
  http://www.nsec.harvard.edu/

Howard University
  http://www.howard.edu/keckcenter/investigators.htm

Hungarian Academy of Sciences Research (Hungary)
  http://www.mfa.kfki.hu/eng/

Institute of Nanotechnology (UK)
  http://www.nano.org.uk/
  http://www.nano.org.uk/books.htm

Iowa State University
  http://www.aere.iastate.edu/
  http://www.public.iastate.edu/~nscentral/news/2007/may/nanotech.
    shtml

James Cook University (Australia)
  http://www.eng.jcu.edu.au/

Korea Institute of Ceramic Engineering and Technology (South Korea)
  http://www.kicet.re.kr/

Lehigh University
  http://www.lehigh.edu/nano/mechanical_nano.html

Ludwig-Mazimillians Universitat Muncgen
http://www.nano.geo.uni-muenchen.de/SW/

Massachusetts Institute of Technology (MIT)
http://dmse.mit.edu/
http://nanoweb.mit.edu
http://www.arl.army.mil/www/default.cfm?page=29
http://snl.mit.edu/

MEXT Nanotechnology Researchers Network Center of Japan
https://nanonet.nims.go.jp/

Michigan State University
http://www.pa.msu.edu/cmp/csc/nanotube.html

Moscow State University
http://polly.phys.msu.ru/

Nanoforum (European Union)
http://www.nanoforum.org/

NanoMechanical Technology Laboratory
http://web.mit.edu/nanolab/facilities.html

Nanotechnology Research Institute (Japan)
http://unit.aist.go.jp/nri/

National University of Singapore (Singapore)
http://www.nusnni.nus.edu.sg/

New York University
http://seemanlab4.chem.nyu.edu/

North Dakota State University
http://www.ndsu.nodak.edu/me/

Northeastern University
http://www.northeastern.edu/chn/

Northwestern University
http://www.mech.northwestern.edu/dept/research/tribology/tribology.
    htm
http://www.iinano.org/
http://www.nsec.northwestern.edu/

Norwegian University of Science and Technology (Norwegian)
  http://www.ntnu.no/kt/english

Oklahoma State University
  http://www.mae.okstate.edu

Paul Scherrer Institute (Switzerland)
  http://lmn.web.psi.ch/

Pennsylvania State University
  http://www.mri.psu.edu/nano/
  http://stm1.chem.psu.edu/
  http://www.gonano.psu.edu/
  http://www.mri.psu.edu/facilities.asp

Princeton University
  www.princeton.edu
  http://www.princeton.edu/~chouweb/

Purdue University
  http://www.physics.purdue.edu/nanophys/
  http://www.physics.purdue.edu/nanophys/newpage10-03/

Rensselaer Polytechnic Institute (NY)
  http://www.rpi.edu/dept/nsec/

Rice University
  http://cnst.rice.edu/
  http://cben.rice.edu//
  http://www.chm.bris.ac.uk/motm/buckyball/c60a.htm

Rutgers University
  http://ceramicmaterials.rutgers.edu/NMSE/tribo.php

Stanford University
  http://mse.stanford.edu/index.html
  http://www.nnun.org/
  http://scpd.stanford.edu/certificates/professional-education-certificate.
    jsp
  http://snf.stanford.edu/About/About.html

Swiss Federal Institute of Technology (Switzerland)
  http://www.ifr.mavt.ethz.ch/
  http://www.met.mat.ethz.ch/

Texas A&M University
  http://www.chem.tamu.edu/cims/
  http://tiims.tamu.edu/director.html

Texas State University
  http://www.txstate.edu/physics/

University of Albany
  http://cnse.albany.edu/

University of Arkansas and University of Oklahoma
  http://www.nhn.ou.edu/cspin/index.html

University of Birmingham (UK)
  http://www.nprl.bham.ac.uk/
  http://nprl.bham.ac.uk/

University of Braunschweig—Institute of Technology (Denmark)
  www.ipat.tu-bs.de

University of California Berkeley
  http://me.berkeley.edu/

University of California Davis
  http://ncnc.engineering.ucdavis.edu/pages/facility.html
  http://ncnc.engineering.ucdavis.edu/

University of California Irvine
  http://www.inrf.uci.edu/index.asp

University of California Los Angeles
  http://www.universityofcalifornia.edu/news/article/5174

University of California Santa Barbara
  http://www.nanotech.ucsb.edu/default.html
  http://engineering.ucsb.edu/news/74
  http://www.cnsi.ucsb.edu/

University of Connecticut
  http://advance.uconn.edu/2008/081027/08102702.htm

University of Copenhagen (Denmark)
  http://nano.ku.dk/english

University of Delaware
  http://www.che.udel.edu/cmet/

University of Florida
http://maic.mse.ufl.edu/
http://www.csb.ufl.edu/

University of Glasgow (UK)
http://www.gla.ac.uk/departments/electronicsandelectricalengineering/
research/micronanotechnology/

University of Illinois Urbana-Champaign
http://facilities.mrl.uiuc.edu/cmm/
http://www.beckman.illinois.edu/index.aspx
http://www.cnst.illinois.edu/

University of Maine
http://www.umaine.edu/lasst/

University of Michigan
http://nano.med.umich.edu/

University of Minnesota
http://www.cems.umn.edu/
http://www.me.umn.edu/~mrz/CNER.htm

University of Missouri–Columbia
http://web.missouri.edu/%7Emae/index.html

University of Moscow (Russia)
http://quantumdot.lanl.gov/klimov.shtml

University of Newcastle, UK
http://research.ncl.ac.uk/nanoscale/

University of Nebraska
http://ascweb.unl.edu/possibilities/mrsec-and-nanotechnology.shtml

University of North Carolina
http://cismm.cs.unc.edu/
http://www.advancedmaterials.unc.edu/links
http://www.physics.unc.edu/~zhou/muri/nccnm.html

University of Notre Dame
http://www.nd.edu/~ndnano/

University of Queensland (Australia)
http://www.aibn.uq.edu.au/

University of Pittsburgh
   http://www.engrng.pitt.edu/mechanical/index.html

University of Southern California
   http://lipari.usc.edu/~lmr/

University of South Carolina
   http://www.cas.sc.edu/phil/scistud/topics.html

University of Southern Mississippi
   http://www.usm.edu/mrsec/

University of Sussex
   http://www.chm.bris.ac.uk/motm/buckyball/c60a.htm
   http://www.nano.sussex.ac.uk/

University of Texas
   http://www.botany.utexas.edu/facstaff/facpages/mbrown/nanopage/

University of ULM (Denmark)
   http://wwwex.physik.uni-ulm.de/nanosnom/

University of Virginia
   http://www.virginia.edu/nanostar/

University of Washington
   http://depts.washington.edu/mse/

University of Western Ontario (Canada)
   http://www.uwo.ca/chem/

University of Wien (Austria)
   http://www.univie.ac.at/spectroscopy/

University of Windsor (Canada)
   http://www.uwindsor.ca/tribology

University of Wisconsin–Madison
   http://mrsec.wisc.edu/
   http://uw.physics.wisc.edu/~himpsel/
   http://www.nanotech.wisc.edu/

Virginia Polytechnic Institute and State University (Virginia Tech)
   http://www.dorn.chem.vt.edu/

Washington State University
  http://www.wsu.edu/Research.html
  http://www.mme.wsu.edu/

Washington University
  http://depts.washington.edu/bionano/index.html
  http://www.nano.washington.edu/index.asp

Widner University
  http://www.science.widener.edu/chemistry/

Yale University
  http://www.seas.yale.edu/
  http://www.eng.yale.edu/reedlab/
  http://www.eng.yale.edu/uelm/nano.html

Nanotechnology and Space

A to Z of Nanotechnology and Nanomaterials, a collaborative venture
  between AZoM and the Institute of Nanotechnology
  http://www.azonano.com/

Autonomous Nanotechnology Swarm Goddard Space Flight and Langley
  Research Center partnership
  http://ants.gsfc.nasa.gov/

Center for Nanospace Technologies, a nonprofit research foundation
  http://www.nanospace.org/

Molecular manufacturing shortcut group, promoting nanotechnology as a
  means to facilitate the settlement of space
  http://www.islandone.org/MMSG/

The Naked Scientists Online, Internet science radio online—sometimes
  covers nanotech
  http://www.thenakedscientists.com/

NanoApex
  http://www.nanoapex.com/

NanoTech-News.Biz, news and events
  http://www.voyle.net/

Nanotechnology.net, news and more, registration required
  http://www.technologynetworks.com/Nano/

Nanotechnology—Now, news of the future sciences and site of the week
  http://www.nanotech-now.com/

Nanotechweb, news, events, and feature articles
  http://nanotechweb.org/

Nano Tsunami, a nonprofit news service
  http://www.nano-tsunami.com/

Nature Materials; also visit the Nanozone (free registration required)
  http://www.nature.com/nmat/index.html
  http://www.nature.com/index.html?g=3&file=/materials/nanozone/
    nanozone.html

NASA Applications of Molecular Nanotechnology, a paper by Al Globus,
  David Bailey, Jie Han, Richard Jaffe, Creon Levit, Ralph Merkle, and
  Deepak Srivastava
  http://www.nas.nasa.gov/Main/Redirector/redirector.php?old_
    url=http://www.nas.nasa.gov/Groups/Nanotechnology/
    publications/

NASA and Self-Replicating Systems: Implications for Nanotechnology, a
  paper by Ralph C. Merkle
  http://www.zyvex.com/nanotech/selfRepNASA.html

News services

Industry

Medical

http://www.nanoindustries.com/links/medical.html
Institute for Bioengineering and Nanoscience in Advanced Medicine
  (IBNAM)
  http://www.ibnam.northwestern.edu/

An Introduction to Nanomedicine at the Jeffline Forum
  http://jeffline.tju.edu/Education/forum/02/10/articles/nano.html

Medical Benefits of Molecular Nanotechnology, a Center for Responsible Nanotechnology paper
http://www.crnano.org/medical.htm

The Nanomedicine Glossary, provided by NanotechNow.com
http://www.nanotech-now.com/nanotechnology-medicine-glossary.htm

Nanomedicine—Information and News at News-Medical-Net
http://www.news-medical.net/search.aspx?q=Nanomedicine

Nanorobots: Medicine of the Future 1999 short paper
http://www.ewh.ieee.org/r10/bombay/news3/page4.html

Nanotechnology and Medicine, by Ralph C. Merkle
http://www.zyvex.com/nanotech/nanotechAndMedicine.html

The National Human Genome Research Institute's Nanomedicine Fact Sheet
http://www.genome.gov/11508736

Robots in the Bloodstream: The Promise of Nanomedicine, by Robert Freitas at KurzweilAI.net
http://www.kurzweilai.net/robots-in-the-bloodstream-the-promise-of-nanomedicine

Tools and Equipment

Accelrys, Inc. (formerly Molecular Simulations, Inc.): Simulation and informatics software.
http://accelrys.com/

Agilent: AFM products to advance nanotechnologies.
http://nano.tm.agilent.com/index.cgi?CONTENT_ID=1666&User:LANGUAGE=en-US

Bid-Service: Used/refurbished scientific and laboratory equipment.
http://www.bidservice.com

CambridgeSoft Corp.: Internet software applications for chemists and engineers.
http://www.cambridgesoft.com//

Capovani Brothers, Inc.: Used equipment science and industry.
http://www.capovani.com/

Cavity Stuffer: Polymer design tool.
  http://www.n-a-n-o.com/nano/cavstuf/cavstuf.html

Crystal Sletchpad: Molecular CAD Tool: Molecular CAD tool to design
  diamondoid nanostructures in full atomic detail.
  http://goanna.cs.rmit.edu.au/~gl/research/nano/crystal.html

DisMol: A freeware molecule display applet.
  http://www.es.embnet.org/Services/MolBio/DisMol/DisMol.html

Equipment for Technology and Science, Inc. (ETS): Preowned
  nanotechnology equipment.
  http://www.equiptechonline.com/

General Nanotechnology: Imaging software for nanomanipulation and
  imaging.
  http://www.gennano.com/

Institute for Microelectronics: Viennese TCAD simulation tools—
  downloadable software.
  http://www.iue.tuwien.ac.at/index.php?id=2

Molecular Visualization Freeware: Protein explorer, chime and rasmol.
  http://www.umass.edu/microbio/rasmol/

Novascan Technologies: AFMs, probes, and control devices.
  http://www.novascan.com/

NT-MDT: Integrated solutions for nanotechnology.
  http://www.ntmdt.com/

The Numerical Alogorithms Group.
  http://www.nag.co.uk/

Physik Instrumente (PI): Nanopositioning, micropositioning, and piezo
  technology.
  http://www.physikinstrumente.com/

Piezo Systems, Inc.: Piezoelectric materials, transducers, and drive
  electronics.

Sci Quip: Refurbished high-tech equipment, many items applicable to
  nanotechnology.
  http://www.sciquip.com/

Veeco Instruments, Inc.: Digital instruments, atomic force, and scanning probe microscopy.
http://www.di.com/

Additional Companies Doing Nanoscale Work

While this list is by no means complete with the companies involved in nanoscale work, it shows the variety of businesses that are currently productive in the field of nanotechnologies. Additional information for investment opportunities, etc., may be on the individual web addresses.
http://www.nanoindustries.com/links/companies.html

3rd Tech Producers of the NanoManipulator DP-100 System: For interactive display and manipulation for nanotechnology research.
http://www.3rdtech.com/

Altair International, Inc. (Nasdaq: ALTI): Plasma spray coating product containing Altair's titania nanoparticles.
http://www.altairtechnologies.com/

Applied Nano Works: Advanced materials experts using next-generation technology to deliver 2–40 nm semiconductor, metal, and oxide crystals produced in water colloids.
http://www.auterrainc.com/

Argonide: Manufacturers of electro-exploded nanosize powders. Participants of the U.S. Nanotechnology Initiative.
http://www.argonide.com/

Bell Labs Innovations.
http://www.alcatel-lucent.com/wps/portal/BellLabs

BioForce Nanosciences, Inc.: Developer of ultra-miniaturized nanoarray technologies for the solid-state, high-throughput analysis of molecular systems.
http://www.bioforcelab.com/

California Molecular Electronics Corporation: Molecular electronics.
http://www.calmec.com/home.html

Cymbet: Manufacturing process for solid-state lithium ion power sources.
http://www.cymbet.com/

Evident Technologies: Manufacturers of EviDots, semiconductor nano-crystals, currently available in Core-CdSe semiconductor nanocrystals offered in a five-color test kit, producing in commercial quantities semi-conductor nanocrystal quantum dots.
http://www.evidenttech.com/

Hewlett Packard HP: Labs working with nanotechnology.
http://www.hpl.hp.com/

Hielscher Ultrasonics: Ultrasound technology development and produc-tion of ultrasonic devices for use in laboratory and industrial applica-tions.
http://www.hielscher.com/ultrasonics/nano_01.htm

Hyperion Catalysis: International producer of carbon nanofiber materials. Hyperion's primary product is Graphite Fibril™ nanotubes.
http://www.hyperioncatalysis.com/

IBM Watson Research Center, Zurich Research Laboratory, Almaden Research Center.
http://www.research.ibm.com/nanoscience/

Isotron Corporation: Nanoparticle composites that improve the physical properties of high-performance industrial coatings.
http://www.isotron.net/

Lightyear Technologies, Inc.: Developer of nanotechnology materials and applications.
http://www.brandmight.com/

Luna Innovations, Inc.: Endohedral metallofullerenes, hollow molecules of carbon atoms that encapsulate various metal and rare earth elements.
http://www.lunainnovations.com/

Mad City Labs: Manufactures nanopositioning systems with subnanome-ter precision.
http://www.madcitylabs.com/

Materials Modification, Inc.: Nanopowder processes in material engineer-ing.
http://www.matmod.com/

Minatec: Center for innovation in micro- and nanotechnology.
http://www.minatec.org/en/

Molecular Manufacturing Enterprises, Inc. (MMEI): Founded to help accelerate advancements in the field of molecular nanotechnology.
http://www.mmei.com/

Nanocor: Nanoclays specifically designed for plastic nanocomposites.
http://www.nanocor.com/

Nanocyl: Producing and commercializing carbon nanotubes of various kinds (multiwall, single wall, functionalized) in bulk quantities.
http://www.nanocyl.com/

Nanogen (NGEN): Microelectronics and molecular biology.
http://www.epochbio.com/

NanoLab: Carbon nanotube-aligned arrays and powder nanotechnology.
http://www.nano-lab.com/

Nanoledge: French supplier of carbon nanotubes.
http://www.nanoledge.com/

Nanologic, Inc.: Integration of nanotechtology into computers.
http://www.nanologicinc.com/

Nanomaterials and Nanofabrication Laboratories (NN-Labs): Selling semiconductor nanocrystals (CdS-CdSe).
http://www.nn-labs.com/

Nanomaterials Research LLC: Nanoengineered materials and devices.
http://www.nrcorp.com/

Nanometrics, Inc. (NANO)
http://www.nanometrics.com/

Nanonex: Offers nanoimprint lithography (NIL) tools, resists, masks, and consulting.
http://www.nanonex.com/

Nanophase: Markets nanocrystalline for commercial applications.
http://www.nanophase.com/

Nanopowders Industries: Special alloy powders for electronic components.
http://www.nanopowders.com/

Nanopower Enterprises, Inc.: Producer of specialty nanopowders.
http://www.nanopowderenterprises.com/

Nanoprobes: Founded to develop the most sensitive reagents and methods for detecting biological molecules.
http://www.nanoprobes.com/

Nanoscale Combinatorial Synthesis, Inc.: Accelerating drug discovery.
http://www.nanosyn.com/

Nanoscale Materials, Inc. (formerly Nantek, Inc.): Materials company developing and commercializing reactive nanoparticles (RNPs) and other related technologies.
http://www.nanoscalecorp.com/

Nanostructured and Amorphous Materials, Inc.: Manufacturer and supplier of varying nanoscale powders.
http://www.nanoamor.com/home

Nanotechnology Systems: Dedicated to the development of ultra-precision machine systems, typically utilizing single-point diamond turning and deterministic micro-grinding technologies, for the production of plano, spherical, aspheric, conformal, and freeform optics.
http://www.nanotechsys.com/

Obducat: Storage, biosensors, and semiconductors.
http://www.obducat.com/

Particular: A start-up company that produces nanoparticles.
http://particular.eu/company.html

PolytecPI, Inc.: Nanopositioning equipment for scanning microscopy.
http://www.polytecpi.com/

SDL Queensgate LTD: Nanopositioning solutions.
http://www.nanopositioning.com/

Sumitomo Electric: Nanoparticles and powders.
http://www.sumitomoelectricusa.com/

Zyvex: Private nanotechnology development company. Goal: To develop and use atomically precise manufacturing to radically change the manufacturing capabilities of the world.
http://www.zyvex.com/

Support Organizations and Professional Societies

ACORN—A Collaboration of Research into Nanoparticles
http://www.globalacorn.com/whatwedo.html

American Institute of Aeronautics and Astronautics (AIAA)
Nanotech 2002
http://www.aiaa.org/content.cfm

American Institute of Chemical Engineers (AIChE)
http://www.aiche.org/annual/

American Society of Mechanical Engineers Nanotechnology Institute
http://www.nanotechnologyinstitute.org/about.html

American Society for Nanomedicine
http://www.amsocnanomed.org/

American Vacuum Society Nanometer Scale S&T Division
http://divisions.avs.org/nstd/default.html

Argonne National Laboratory (ANL)
http://www.anl.gov/

Asia Nano Forum (ANF) Network—supported by 13 economies in the Asia
Pacific region
http://www.asia-anf.org/

Asian Technology Information Program
http://www.atip.org/

Australia Commonwealth Scientific and Industrial Research Organization
(CSIRO)
http://www.csiro.au/org/CMSE.html

Basic Energy Sciences (BES)—Nanoscale Science, Engineering and
Technology Research
http://www.er.doe.gov/bes/reports/abstracts.html

Basic Energy Sciences (BES)—Materials Chemistry and Bimolecular
Materials
http://scgf.orau.gov/BES.html

Brookhaven National Laboratory (BNL) Center for Functional
Nanomaterials
http://www.bnl.gov/cfn/

The Center for Computational Sciences at the ORNL (Oak Ridge National
Laboratory)
http://www.nccs.gov/

The Center for Engineering Science Advanced Research at the ORNL
  http://www.cesar.ornl.gov/

The Center for Nanophase Materials Sciences at the ORNL
  http://www.cnms.ornl.gov/

Chesapeake Nanotech Initiative (CNI)
  http://www.nano.gov/html/meetings/srw2005/

Chicago Microtechnology/Nanotechnology Community (CMNC)
  http://www.nlake.com/nanotech.htm

Colorado Nano Technology Initiative
  http://www.coloradonanotechnology.org/home/

Connecticut Nanotech Initiative
  http://www.nanotech-now.com/news.cgi?story_id=21223

European Nanoelectronics Initiative Advisory Council
  http://www.eniac.eu/

Foreign Ministry of Education and Research (BMBF)
  http://www.bmbf.de/en/

GDRE NanoE—GDR on Science and Applications of Nanotubes
  http://www.graphene-nanotubes.org/

Greater Garden State Nanotechnology Alliance (GGSNA)
  http://www.njtc.org/community/industry/ggsna.asp

Idaho National Engineering and Environmental Laboratory (INEEL)
  https://inlportal.inl.gov/portal/server.pt/community/about_inl/259/
    capabilities/1511

I Love New York Nanotech
  http://www.nylovesnano.com/

India Yashnanotech
  http://www.yashnanotech.com/

Institute of Electrical and Electronics Engineers, Inc. (IEEE)
  Nanotechnology Council; Virtual Communities
  http://ewh.ieee.org/tc/nanotech/
  http://ewh.ieee.org/soc/pcs/

Institute for Soldier Nanotechnologies
  http://web.mit.edu/isn/

Invest Australia
  http://www.austrade.gov.au/Buy/default.aspx

Israel National Nanotechnology Initiative (INNI)
  http://www.nanoisrael.org/default.asp

Italian Center for Nanotechnology, branch of AIRI (Italian Association for
  Industrial Research)
  http://www.nanotec.it/eng/index_eng.html

Japan Federation of Economic Organizations
  www.keidanren.or.jp/

Japan Frontier Research System (FRS) RIKEN
  http://www.riken.jp/engn/r-world/research/lab/asi/index.html

Lawrence Berkeley National Laboratory (LBNL)
  http://www.lbl.gov/

Lawrence Livermore National Laboratory (LLNL)
  http://www.anl.gov

Los Alamos National Laboratory (LANL)—Nanoscience and Technology
  http://www.lanl.gov/mst/nano/

MANA: Mid-Atlantic Nanotechnology Alliance (Tristate PA-NJ-DE)
  http://www.midatlanticnano.org/

Materials Research Society
  http://www.mrs.org/s_mrs/alias.asp

MEXT Nanotechnology Researchers Network Center of Japan
  http://www.nanonet.go.jp/english/

Minnesota Nanotechnology Initiative (MNI)
  http://thor.ece.umn.edu/

The Molecular Foundry at LBNL
  http://foundry.lbl.gov/

NaNet—Danish Nanotechnology Network
  http://www.dtu.dk/Centre/NaNet/English.aspx

NanoCMOS
  www.nanocmos.ac.uk

Nano-FIB Project
  http://www.nanofib.com

Nano-Map
  http://www.nano-map.de/index.php?mode=1&lang=en#start_
    BRDNational Center for

Nanomaterials Laboratories (NML) Japan
  http://www.nims.go.jp/nanomat_lab/greeting.htm

NanoMEMS Edmonton
  http://www.nsti.org/Nanotech2005/exhibitor.html?id=147

Nano Øresund
  http://www.nano-oresund.org/web/default.asp

Nano Quebec
  http://www.nanoquebec.ca/nanoquebec_w/site/index.
    jsp?NOPRESERVElanguageID=0

NanoRoadMap (NRM)
  http://www.nanoroadmap.it/

Nanoscale Integrated Processing of Self-Organizing Multifunctional
  Organic Materials (NAIMO)
  http://www.ec.europa.eu/research/fp6/projects.cfm?p=3

Nanoscience and Technology (NCNST) of China
  http://english.nanoctr.cas.cn/

Nanotechnology Association of Ireland
  http://nanotechireland.com/

Nanotechnology Issues Dialogue Group (NIDG)
  http://webarchive.nationalarchives.gov.uk/+/http://www.dius.
    gov.uk/office_for_science/science_in_government/key_issues/
    nanotechnologies/nidgNanotechnology and Nanoscience
  http://www.azonanp.com/suppliers.asp?SupplierID=168

Nanotechnology Technical Advisory Group (NTAG)
  http://www.nema.org/stds/international/iec-TAGs/tc113.cfm

Nanotechnolog—UK Trade and Investments in the United States
http://ukinusa.fco.gov.uk/en/?Sarticletype=24&other_ID=334

Nano2Life (N2L)
http://www.nanno2life.org

NASA
http://spacebiosciences.arc.nasa.gov/

NASA Ames Center for Nanotechnology
http://www.ipt.arc.nasa.gov/

NASA Ames Research Center
http://www.nasa.gov/centers/ames/home/index.html

NASA Ames Research Center, Center of Nanotechnology
http://www.ipt.arc.nasa.gov/Graphics/arl_talk.pdf

NASA Glenn Research Center
http://www.nasa.gov/centers/glenn/home/index.html

NASA-JSC Area Nanotechnology Study Group
http://www.nanonewsboard.com/

NASA and JSC Carbon Nanotube Project
http://mmptdpublic.jsc.nasa.gov/jscnano/

NASA Michigan State University Clemson U of P CMU—Use of Carbon
Nanotubes in Space
http://www.pa.msu.edu/cmp/csc/nasa/

National Engineering Research Center for Nanotechnology (NERCN)
www.rusnano.com/Admin/Files/FileDownload.aspx?id=2998

National Institute for Materials Science (NIMS), Japan
http://www.nims.go.jp/eng/

National Institute of Standards and Technology (NIST)
http://www.nist.gov/index.html

National Nanofabrication Users Network (NNUN)
http://www.nnun.org/

National Nanotechnology Centre, Thailand
http://www.nanotec.or.th/en/

National Nanotechnology Coordination Office (NNCO)
  http://www.nano.gov/html/about/nnco.html

National Nanotechnology Initiative (NNI), U.S. federal government
  program, NNI Strategic Plan
  http://www.nanolawreport.com/2010/07/articles/nni-strategic-plan-
    2010-request-for-information/

National Science and Technology Council (NSTC)
  http://www.whitehouse.gov/administration/eop/ostp/nstc

National Science Foundation (NSF)
  http://www.nsf.gov/

Natural Sciences and Engineering Research Council of Canada (NSERC)
  http://www.nserc-crsng.gc.ca/index_eng.asp

NCI Alliance for Nanotechnology in Cancer
  http://nano.cancer.gov/

NCI Nanotechnology Characterization Lab—National Institute of
  Standards and Technology (NIST), U.S. Food and Drug Administration
  (FDA), National Cancer Institute (NCI)
  http://ncl.cancer.gov/

New Jersey Nanotechnology Consortium (NJNC)
  http://www.njnano.org/

NIH Nanotechnology and Nanoscience Information
  http://www.nibib.nih.gov/Research/NIHNano

NIH National Human Genome Research Institute Nanomedicine Fact Sheet
  http://www.genome.gov/11508736

NIST—Advanced Technology Program
  http://www.atp.nist.gov/#

NIST—Center for Nanoscale Science and Technology
  http://www.nist.gov/cnst/

NIST—Combinatorial Methods Center
  http://www.nist.gov/mml/polymers/combi.cfm

NIST—Nanoelectronic Device Metrology
  http://www.nist.gov/pml/semiconductor/cmos/nedm.cfm

North Carolina Nanotechnology Initiative
http://www.ncnanotechnology.com/public/root/news.asp

Northern California Nanotechnology Initiative (NCnano)
http://www.ncnano.org/

Northwest Nanoscience and Nanotechnology Network
http://www.pnl.gov/nano/

Oklahoma Nanotechnology Initiative
http://www.oknano.com/

Oregon Nanotechnology Initiative
http://www.pnl.gov/nano/

President's Council of Advisors on Science and Technology (PCAST)
http://www.whitehouse.gov/administration/eop/ostp/pcast

Sandia National Laboratories (SNL)
http://www.sandia.gov/

Science and Technology Policy Institute (STPI)
https://www.ida.org/stpi.php

Shanghai Nanotechnology Promotion Center
http://www.snpc.org.cn/english/testing.asp

SINANO
http://www.sinano.eu/

South African Nanotechnology Initiative (SANi)
http://www.sani.org.za/

South Carolina Nanotechnology Initiative
http://www.areadevelopment.com/specialPub/southernTech08/
national-nanotechnology-initiative-southern-states.shtml

State initiatives (United States)
http://www.nano.gov/041805Initiatives.pdf

Texas Nanotechnology Initiative
http://www.nanotx.biz/

UK Micro and Nanotechnology Network
http://www.ktn.innovateuk.org/welcome

UK National Contact Point for Nanotechnologies, Materials and Production
    Processes (NMP)
    http://www.fp7uk.dti.gov.uk/Site/ThematicAreas/Nanosciences/
        default.cfm

U.S. Food and Drug Administration
    http://www.understandingnano.com/nanotechnology-regulation.html

U.S. government nanotechnology programs searchable database
    http://cint.lanl.gov/

U.S. Naval Research Laboratory
    http://cst-www.nrl.navy.mil/ResearchAreas/Nanostructures/
    http://www.nrl.navy.mil/nanoscience/

Veneto Nanotech
    http://www.venetonanotech.it/en/

Virginia Nanotechnology Initiative (VNI)
    http://www.nano.gov/html/about/symposia.html

Washington Nanotechnology Initiative
    http://www.watechcenter.org/?p=Washington+Nanotechnology
        +Initiative&s=478

Nonprofit Organizations

The American Vacuum Society: Science and technology of materials,
    interfacing, and processes.
    http://www.avs.org/

Center for Responsible Nanotechnology: Awareness of benefits and
    dangers of nanotechnology.
    http://www.crnano.org/

The Foresight Institute: The nonprofit definitive source of knowledge
    on nanotechnology. Provides yearly conferences and other events that
    include a wide span of nanotechnology topics.
    http://www.foresight.org

IEEE Nanotechnology: Virtual community, a professional society.
    http://ieeenano.mindcruiser.com/

The Institute of Nanotechnology (UK): Registered charity promoting a
    focus on nanotechnology.
    http://www.nano.org.uk/

Larta: Reports, consulting, and training on commercialism.
http://www.larta.org/

NanoSig: Nonprofit organization connecting investors, entrepreneurs, and
large corporations with deal flow, education, information, and IP via
conferences, seminars, and global networking.
http://www.nanosig.org/

The Scripps Research Institute: Private research
http://www.scripps.edu/e_index.html

Publications

Books

The Age of Spiritual Machines: When Computers Exceed Human Intelligence, by
Ray Kurzweil

*Asianano 2002: Proceedings of the Asian Symposium on Nanotechnology and
Nanoscience,* edited by Masatsugu Shimomura, Teruya Ishihara

Becoming Immortal: Nanotechnology, You, and the Demise of Death, by Wesley
M. Du Charme.

By the Light of the Moon, by Dean Koontz (fiction)

Carbon Nanotubes, edited by M. Endo, M.S. Dresselhaus, S. Iijima

Designing the Molecular World, by Philip Ball

The Diamond Age: Or, Young Ladies Illustrated Primer, by Neal Stephenson
(fiction)

Discovering the Nanoscale, edited by Davis Baird, Alfred Nordmann, and
Joachim Schummer

Encyclopedia of Nanoscience and Nanotechnology, 10-volume set with forward
by Richard Smalley

Engines of Creation, by K. Eric Drexler

Fantastic Voyage, by Ray Kurzweil and Terry Grossman, MD

Frankenstein: Lost Souls, by Dean Koontz (fiction)

From Instrumentation to Nanotechnology, by J.W. Gardner and H.T. Hingle

*The Global Technology Revolution: Bio/Nano/Material Trends and Their Synergies
with Information Technology,* by Philip S. Anton, Richard Silberglitt, and
James Schneider (Rand Corp.)

Great Mambo Chicken and the Transhuman Condition, by Ed Regis

Handbook of Micro/Nanotribology (Mechanics and Materials Science), by Bharat
Bhushan (editor)

Handbook of Nanoscience, Engineering, and Technology, by Donald Brenner,
Sergey Lyshevski, Gerald Iafrate, and William A. Goddard III

Handbook of Nanostructured Materials and Nanotechnology, Volumes 1-5, by Hari Singh Nalwa

Handbook of Nanotechnology Business Policy, and Intellectual Property Law, by John C. Miller, Ruben Serrato, Jose Miguel Represas-Cardenas, and Griffith Kundahl

Integrated Chemical Systems: A Chemical Approach to Nanotechnology, by Allen J. Bard

Introduction to Nanotechnology, by Charles P. Poole

Investing in Nanotechnology: Think Small, Win Big, by Jack Uldrich

The Investor's Guide to Nanotechnology and Micromachines, by Glenn Fishbine

Leaping the Abyss: Putting Group Genius to Work, by Gayle Pergamit and Chris Peterson

Micromachines: A New Era in Mechanical Engineering, by Iwao Fujimasa

Micromachines and Nanotechnology: The Amazing New World of the Ultrasmall, by David Darling

Molecular Engineering of Nanosystems, by Ed Rietman

Nano: The Emerging Science of Nanotechnology, Vol. 1, by Ed Regis

Nano and Microelectromechanical Systems: Fundamentals of Nano- and Microengineering, by Sergey Edward Lyshevski

Nanofabrications and Biosystems: Integrating Materials Science, Engineering and Biology, by Harvey C. Hoch, Harold G. Craighead, and Lynn Jelinski

Nanomaterials: Synthesis, Properties and Applications, by A.S. Edelstein and R.C. Cammarata

Nanomedicine: Basic Capabilities, Vol. 1, by Robert Freitas Jr.

Nanomedicine, Vol. 2, Biocompatibility, by Robert A. Freitas Jr.

Nanophysics and Nanotechnology: An Introduction to Modern Concepts in Nanoscience, by Edward L. Wolf

Nanoscale Assembly: Chemical Techniques Edition, by W.T. Huck and Wilhelm T.S. Huck

Nanosystems: Molecular Machinery, Manufacturing and Computation, by K. Eric Drexler

Nanotechnology by Gregory Timp (editor)

Nanotechnology: A Gentle Introduction to the Next Big Idea, by Mark A. Ratner and Daniel Ratner

Nanotechnology: An Introduction to Nanostructuring Techniques, by Michael Kohler

Nanotechnology: Basic Science and Emerging Technologies, by Michael Wilson

Nanotechnology: Molecular Speculations on Global Abundance, edited by B.C. Crandall

Nanotechnology Research Directions: IWGN Workshop Report—Vision for Nanotechnology in the Next Decade, by IWGN Workshop, R.S. Williams, P. Alivisatosm, and Mihail C. Roco

Nanotechnology: Risks, Ethics and Law, edited by Geoffrey Hunt and Michael Mehta

Nanotechnology: Science, Innovation and Opportunity, by Lynne E. Foster, forewords by George Allen and Joe Lieberman

Nanotechnology: Towards a Molecular Construction Kit, by Arthur ten Wolde

Nanotechnology for Dummies, by Earl Boysen and Richard D. Booker

Nanotechnology in Catalysis, vol. 2, edited by Bing Zhou, by Gabor Somorjai and Sophie Hermans

Nanotechnology in Construction, edited by W. Zhu, P. J. M. Bartos, J. J. Hughes, and P. Trik

Nanotechnology and the Environment, by Barbara Karn, Paul Alivasatos, Vicki Colvin, and Wei-Xian Zhang

Nanotechnology Global Strategies, Industrial Trends and Applications, edited by Jurgen Schulte

Nanotechnology Molecularly Designed Materials, edited by Gan-Moog Chow and Kenneth Gonsalves

Nanotechnology and Nano-Interface Controlled Electronic Devices, edited by M. Iwamoto, K. Kaneto, and S. Mashiko

Nanotechnology Playhouse, by Christopher Lampton

The Next Big Thing Is Really Small: How Nanotechnology Will Change the Future of Your Business, by Jack Uldrich and Deb Newberry

Our Molecular Future: How Nanotechnology, Robotics, Genetics and Artificial Intelligence Will Transform Our World, by Douglas Mulhall

Prey, by Michael Crichton (fiction)

Prospects in Nanotechnology: Toward Molecular Manufacturing, by Markus Krummenacker and James Lewis

Recent Advances and Issues in Molecular Nanotechnology, by David E. Newton

Soft Machines: Nanotechnology and Life, by Richard A.L. Jones

TechTV's Catalog of Tomorrow, by Andrew Zolli (editor)

There's Plenty of Room at the Bottom, by Richard Feynman

Thin-Film Solar Cells: Next Generation Photovoltaics and Its Applications, by Yoshihiro Hanakawa

Travels to the Nanoworld: Miniature Machinery in Nature and Technology, by Michael Gross

Unbounding the Future, by K. Eric Drexler, Chris Peterson, and with Gayle Pergamit

Understanding Nanotechnology, by Scientific American

What Is Nanotechnology and Why Does It Matter: From Science to Ethics, by Fritz Allhoff, Patrick Lin, and Daniel Moore

Research Publications and Magazines

Biomedical Microdevices
http://iopscience.iop.org/0957-4484

IEEE Nanotechnology Magazine
http://ieeexplore.ieee.org/xpl/RecentIssue.jsp?reload=true&punum
ber=4451717

Institute of Physics Nanotechnology
   http://iopscience.iop.org/0957-4484

Journal of Nanoscience and Nanotechnology, American Scientific Publishers
   http://www.aspbs.com/jnn/

Journal of Nanoparticle Research
   http://www.springerlink.com/content/103348/

MIT's Technology Review
   http://www.technologyreview.com/Nanotech/

Nano Letters American Chemical Society Publications and the Institute of
   Physics
   http://pubs.acs.org/journal/nalefd
   http://iopscience.iop.org/0957-4484

NASA Tech Briefs
   http://www.techbriefs.com/

Virtual Journal of Nanoscale Science and Technology
   http://www.vjnano.org/nano/Nanotechnology Streaming Videos, online
      and free to download
   http://particular.eu/video.html

References

Abrams, R. What does it take to impress an angel investor? Small Business and Small Business Information for Entrepreneurs. Retrieved March 2011 from http://www.inc.com/articles/2001/03/22375.html.

Aeppli, G., and Broholm C. 1994. In *Handbook on the Physics and Chemistry of Rare-Earths*, ed. K.A. Gschneidner, 123–175. Vol. 19. Amsterdam: North-Holland Elsevier.

Aitken, R., Butz, T., Colvin, V., Maynard, A., et al. 2006. *Safe Handling of Nanotechnology* 444(16): 267–269.

Akesson, A. Hedge Fund Moves: German Nanotech Investor Prepares for ADR Trading in the US. Hedge Fund News HedgeCo.Net. Retrieved March 2011 from http://www.hedgeco.net/news/10/2010/hedge-fund-moves-german-nanotech-investor--prepares-for-adr-trading-in-the-us.html.

Akesson, A. Nanotec Hedge Funds Soar in 2010. Hedge Fund News HedgeCo.Net. Retrieved March 2011 from http://www.hedgeco.net/news/09/2010/nanotec-hedge-fund-soars-in-2010.html.

Anderson, P., and Tushman, M. L. 1990. Technological Discontinuities and Dominant Designs: A Cyclical Model of Technological Change. *Administrative Science Quarterly* 35:604–633.

ANGLE Technology Group. 2004. Commonwealth of Pennsylvania Nanotechnology Strategy. August 10.

ANSI Nanotechnology Standards Panel. Retrieved March 19, 2010, from http://www.ansi.org/standards_activities/standards_boards_panels/nsp/overview.aspx?menuid=3.

Aragon, L. 2004. Nano-Bubble Ahead? *Venture Capital Journal*, July 1, p. 1.

Arnall, A.H. 2003. *Future Technologies, Today's Choices*. Canonbury Villas, London: Greenpeace Environmental Trust.

ASTM International Standards Worldwide. Retrieved March 18, 2010, from http://www.astm.org/ABOUT/aboutASTM.html.

AUTM. 2003. *Licensing Survey: FY 2002*. A.J. Stevens, Association of University Technology Transfer Managers.

Bakand, S., et al. 2005. Toxicity assessment of industrial chemicals and airborne contaminants: Transition from *In Vivo* to *In Vitro* Test Methods: A Review. *Inhalation Toxicology*.

Bastani, B., and Fernandez. 2001. Intellectual Property Rights in Nanotechnology. Fernandez & Associates, LLP. Retrieved March 2011 from www.iploft.com.

Berger, M. Legal Implications of the Nanotechnology Patent Land Rush. Retrieved March 2011 from http://www.nanowerk.com/spotlight/spotid=1919.phps.

Bergman, B. 2005. License to grow: An explosion of patent filings brings with it an increase in litigation cases. What can a savvy investor do? *Venture Capital Journal* 1, April 1.

Blochl, P.E., Joachim, C., and Fisher, A.J. 1993. *Computations for the Nanoscale*. Kluwer Academic.

Braunschweig, C. 2003. Nano nonsense: Venture capitalists are searching for the next big thing in the smallest of technologies. But even with the most powerful electron microscope, they will find it very difficult to detect a profit. *Venture Capital Journal* 1, January 1.

Braunschweig, C. 2004. Nanotech Company Raises $45M Recap. *Private Equity Week* 11(21), 2.

Brindell, J.R. 2009. Nanotechnology and the dilemmas facing business and government. *Florida Bar Journal* 83(7): 73.

Brookstein, D. 2005. *Nanotech Fortunes—Make Yours in the Boom*. San Diego: The Nanotech Company.

Brown, P.B. Seven Steps to Heaven. Retrieved March 2011 from http://www.inc.com/magazine/20011001/23478.html.

Bruns, B. 2001. Open sourcing nanotechnology research and development: Issues and opportunities. *IOP Electronic Journals* 12(3): 198–210.

Burri, R.V., et al. 2007. Public perception of nanotechnology. *Journal of Nanoparticle Research*.

Chemical Week 167(42): 22–23, 2005.

Chen, Z., et al. 2006. Acute toxicological effects of copper nanoparticles *In Vivo*. *Toxicology Letters*.

Coase, R.H. 1990. *The Firm, the Market and the Law*. Chicago: University of Chicago Press.

Council for Science and Technology. 2007a. Nanosciences and Nanotechnologies. http://www2.cst.gov.uk/cst/business/files/nano_review.pdf.

Council for Science and Technology. 2007b. Nanotechnologies Review. http://www2.cst.gov.uk/cst/business/nanoreview.shtml.

Crandall, B.C., ed. 1996. *Nanotechnology: Molecular Speculations on Global Abundance*. Cambridge, MA: MIT Press.

Dahl, D. Earning Your Wings. Retrieved March 2011 from http://www.inc.com/magazine/20050101/getting-started.html.

Darby, M., and Zucker, L. 2003. *Grilichesian Breakthroughs: Inventions of Methods of Inventing and Firm Entry in Nanotechnology*. National Bureau of Economic Research.

Davey, M.E. 2006. *Manipulating Molecules: Federal Support for Nanotechnology Research*. CRS Report for Congress.

Davies, J.C. 2006. Stricter nanotechnology laws are urged. *Washington Times*, January 11, p. A02.

Davis, S., and Davidson, B. 1991. *2020 Vision: Transform Your Business Today to Succeed in Tomorrow's Economy*. New York: Simon and Schuster.

Delevett, P. Venture Capital Investment in 2010 Grew for First Time in Three Years. Retrieved March 2011 from http://www.mercurynews.com/business/ci_17443791?nclick_check=1.

Department of Energy (DOE). 2007. Approach to Nanomaterial ES&H. Retrieved September 22, 2010, from http://www.sc.doe.gov/bes/DOE_NSRC_Approach_to_Nanomaterial_ESH.pdf.

Department of Health and Human Services. 2007. *Progress Toward Safe Nanotechnology in the Workplace: A Report from the NIOSH Nanotechnology Research Center*. CDC Workplace Safety and Health.

Dixon, J. 2007. The Diversity of Natural Nanoparticles in Soils. Presented at Symposium on Characterization and Reactivity of Natural and Synthetic Nanoparticles in Soils.

Doing Business 2010. The World Bank. Ranking on 1,832 countries. Retrieved November 2010 from http://www.doingbusiness.org/economyrankings/.

Drews, D., Ehrfeld, W., Lacher, M., Mayr, K., Noell, W., Schmitt, S., and Abraham, M. 1999. Nanostructured probes for scanning near-field optical microscopy. *Nanotechnology* 10: 61-64.

Drexler, K.E. 1986. *Engines of Creation*. New York: Anchor Books. Available at www. foresight.org.

Drexler, K.E., and Peterson, C., with Pergamit, G. 1991. *Unbounding the Future: The Nanotechnology Revolution*. New York: William Morrow.

DuCharme, W.M. 1995. *Becoming Immortal: Nanotechnology, You and the Demise of Death*. Evergreen, CO: Blue Creek.

Elechiguerra, J.L., et al. 2005. Interaction of silver nanoparticles with HIV-1. *Journal of Nanotechnology*.

European Nanotechnology Trade Alliance (ENTA). http://www.euronanotrade. com/.

Environment Protection Agency. 2007. Nanotechnology White Paper. http://es.epa. gov/ncer/nano/publications/whitepaper12022005.pdf.

Environment Protection Agency. Nease Chemical Superfund Site. http://www.epa. gov/region5/sites/nease/.

Feynman, R.P. 1960. There's Plenty of Room at the Bottom: An Invitation to Enter a New Field of Physics. *Engineering and Science*. www.zyvex.com/nanotech/ feynman.html.

Fishbine, G. 2002. *The Investor's Guide to Nanotechnology and Micromachines*. New York: John Wiley & Sons.

Fluckiger, S. 2006. Industry's challenge to academia: Changing the bench to bedside paradigm. *Experiential Biology and Medicine*.

Foley, E.T., and Hersam, M.C. 2006. Assessing the need for nanotechnology education reform in the United States. *Nanotechnology Law and Business* 3(4): 467–484.

Food and Drug Administration (FDA). FDA Regulation of Nanotechnology Products. http://www.fda.gov/nanotechnology/regulation.html.

Forbes. Venture Capital Fundraising Continues Its Downward Slide. Retrieved January 20, 2011, from http://blogs.forbes.com/maureenfarrell/2011/01/18/venture-capital-fundraising-continues-its-downward-slide/.

Forbes Nanotechnology. Retrieved January 11, 2011.

Franco, A., et al. 2007. Limits and prospects of the "Incremental Approach" and the European legislation on the management of risks related to nanomaterials. *Regulatory Toxicology and Pharmacology*.

Freitas, Robert A., Jr. 1999. *Nanomedicine: Basic Capabilities*. Vol. I. Austin, Texas: Landes Bioscience.

French, S.F. 2000. Common Interest Communities: The dilemma of shared resources in residential housing. *Common Property Resource Digest* (51): 4-5.

Friedman, R.S., McAlpine, M.C., Ricketts, D.S., Ham, D., and Lieber, C.M. 2005. high-speed integrated nanowire circuits. *Nature* 434: 1085.

Gauthier-Jaques, A. 2005. Financial Market Perspective—The Commercial Potential of Nanotechnology and the Value for Investors. *Journal of Medical Marketing* 5: 36.

Geiger, R.L., and Hallacher, P. 2005. *Nanotechnology and the States Public Policy*. University Research and Economic Development in Pennsylvania, Pennsylvania State University.

Geisler, E. 1995. Industry-University Technology Cooperation: A Theory of Interorganizational Relationships. *Technology Analysis and Strategic Management* 7(2): 217–229.

Griffitt, R.J., et al. 2007. Exposure to copper nanoparticles causes gill injury and acute lethality in zebrafish. *Environmental Science and Technology*.

Gubrud, M.A. 1997. Nanotechnology and International Security. Paper presented at the Fifth Foresight Conference on Molecular Nanotechnology, Palo Alto, CA, November 5–8.

Gupta, U. 2001. For venture capitalists, life sciences is hot. *Venture Capital Journal*, September 1.

Halperin, J.L. 1998. *The First Immortal*. New York: Del Rey.

Hanson, R. 1998. A Critical Discussion of Vinge's Singularity Concept. www.extropy.org/eo/articles/vi.html.

Hassan, E., and Sheehan, J. 2003. *Scaling-Up Nanotechnology*. Organization for Economic Cooperation and Development.

Hawken, P., Lovins, A.B., and Hunter Lovins, L. 1999. *Natural Capitalism: The Next Industrial Revolution*. London: Earthscan.

Heller, J., and Peterson, C. Valley of Death in Nanotechnology Investing. Retrieved April 2011 from http://www.foresight.org/policy/brief8.html.

Hodge, G., Bowman, D., and Binks, P. 2005. Nanotechnology, the next big business. *Monash Business Review* 1(2): 1–11.

Holman, M., Kemsley, Nordan, M., et al. 2006. *The Nanotech Report*. 4th ed. Investment Overview and Market Research for Nanotechnology.

Horsley, P. 1997. *Trends in Private Equity*. National Venture Capital Association.

Horton, M.A. 2000, Integrating Confocal Microsopy with an Atomic Force Microscope. *Single Molecules*.

Institute of Occupational Medicine (IOM). http://www.iom-world.org/.

International Electrotechnical Commission (IEC). From Lab to Real Life. Retrieved March 13, 2010, from http://www.iec.ch/online_news/etech/arch_2010/etech_0110/tc_3.htm.

Japan Nanotechnology Risk and Standardization Efforts. Retrieved March 13, 2010, from http://unit.aist.go.jp/nanotech/apnw/articles/library3/pdf/3-39.pdf.

Ikezawa, N. 2001. *Nanotechnology: Encounters of Atoms, Bits, and Genomes*. NRI Papers, No. 62.

Ikezawa, N. 2002. *Nanotechnology to Japan's Rescue*. Kodansha.

Ikezawa, N. 2004. *The Role of Venture Businesses in Supporting the Commercialization of Nanotechnology*. Vol. 7. Nomura Research Institute.

Journal of the Minerals, Metals, and Materials. 2004. Venture Capital—Corporations Invest in Nanotechnology's Future. *Journal of the Minerals, Metals, and Materials*, 6.

Joy, B. 2000. Why the Future Doesn't Need Us. *Wired Magazine* 8.04, April 2000. Retrieved June 16, 2010, from www.wired.com/wired/archive/8.04/joy.html.

Kai, H.O., Pelkonen et al. 2003. Accumulation of Silver from Drinking Water into Cerebellum and *Musculus soleus* in Mice. *Toxicology*.

Kantrowitz, A. 1992. The weapon of openness. In *Nanotechnology Research and Perspectives*, ed. B.C. Crandall and J. Lewis. Cambridge, MA: MIT Press. Retrieved June 6, 2010, from http://www.foresight.org/Updates/Background4.html.

Knol, E. 2004. Micro and nanotechnology commercialization: Balance between exploration and exploitation. In *MANCEF COMS2004 Conference*, pp. 215–220.

Krazit, T. 2005. HP Technology Lets Future Chips Live with Mistakes. *InfoWorld Magazine*, June 9.

Kurzweil, R. 1999. *The Age of Spiritual Machines*. New York: Viking.

Kurzweil, R. 2005. *The Singularity Is Near: When Humans Transcend Biology*. New York: Viking.

Kurzweil, R. 2006. *Nanotechnology Dangers and Defenses Nanotechnology Perceptions: A Review of Ultraprecision Engineering and Nanotechnology.* Vol. 2, No. 1, March 27. http://www.kurzweilai.net/meme/frame.html?main=/articles/art0653.html.

Lavinsky, D. When, Where, and How to Raise Venture Capital. Retrieved March 2011 from http://www.businessinsider.com/when-where-and-how-to-raise-venture-capital-2011-3.

Lawrence, S. 2005. Nanotech Grows Up. *Technology Review,* June, p. 31.

Lawton, J. Making Friends: The Name of the Angel Game. Retrieved April 2011 from http://www.inc.com/articles/2000/02/19502.html.

Lee, A. Examining the Viability of Patent Pools to the Growing Nanotechnology Patent Thicket. University of Virginia.

Lehenkari, P.P., Charras, G.T., Nesbitt, S.A., and Horton, M.A. 2000. Adapting atomic force microscopy for cell biology. *Ultramicroscopy* 82: 289–295.

Lighthill Risk Network. http://www.lighthillrisknetwork.org/.

Linquiti, P., and Teepe, P. 2006. *Characterizing the Environment, Health, and Safety Implications of Nanotechnology: Where Should the Federal Government Go from Here?* ICF International.

Lux Research. 2006. *The Nanotech Report.* 4th ed.

Lux Research. Retrieved June 6, 2010, from http://www.luxresearchinc.com/.

Mace, C., et al. 2006. Nanotechnology and groundwater remediation: A step forward in technology understanding. *Remediation Journal.*

Macoubrie, J. 2005. Informed Public Perceptions of Nanotechnology and Trust in Government. The Project on Emerging Technologies.

Masnick, M. Next Tech Area to Be Hindered by Patents: Nanotech … and Much of It Is Funded with Your Tax Dollars. Techdirt. Retrieved February 2011 from http://www.techdirt.com/articles/20110211/21345813068/next-tech-area-to-be-hindered-patent-nanotech-much-it-is-funded-with-your-tax-dollars.shtml.

Maynard, A. 2006a. Is Nanotechnology Hazardous to Your Health? *Machine Design* 78(23): 71.

Maynard, A. 2006b. *Research on Environmental and Safety Impacts of Nanotechnology: What Are the Federal Agencies Doing?* U.S. House of Representatives Committee on Science.

Mazzola, L. 2003. Commercializing nanotechnology. *Nature Biotechnology* 21(10): 1–7.

Monteiro-Riviere, N.A., et al. 2006. Challenges for Assessing Carbon Nanomaterial Toxicity to the Skin. *Carbon.*

Muller, J., et al. 2005. Respiratory Toxicity of Multi-Wall Carbon Nanotubes. *Toxicology and Applied Pharmacology.*

National Institute of Advanced Industrial Science and Technology (AIST). http://www.aist.go.jp/index_en.html.

Nanobiotechnology: Report of the National Nanotechnology Initiative Workshop, October 9–11, 2003.

NanoMaterials Commercialism—PA NanoMaterials Commercialization Center. Retrieved November 2010 from http://www.pananocenter.org/nano-center-about.aspx.

Nano Risk Framework. http://nanoriskframework.com/.

Nanoscale Science, Engineering, and Technology Subcommittee. 2005a. Nanoscience Research for Energy Needs: Report of the National Nanotechnology Initiative Grand Challenge Workshop, March 16–18, 2004.

Nanoscale Science, Engineering, and Technology Subcommittee. 2005b. Nanotechnology: Societal Implications—Maximizing Benefits for Humanity Report of the National Nanotechnology Initiative Workshop, December 2–3, 2003.

Nanoscale Science, Engineering, and Technology Subcommittee. 2005c. Regional, State, and Local Initiatives in Nanotechnology: Report of the National Nano-technology Initiative Workshop, September 30–October 1, 2003.

Nanoscale Science, Engineering, and Technology Subcommittee. 2006. Instrumentation and Metrology for Nanotechnology: Report of the National.

Nanoscience and Nanotechnology: Opportunities and Challenges in California. 2004. California Council on Science and Technology.Nano Tech. 2010. International Exhibition and Conference. Press release, March 13. http://www.nanotechexpo.jp/en/pdf/pressrelease100311_e.pdf.

Nanotechnology Breakthroughs of the Next 15 Years. 2007. *Futurist* 41: 4.

Nanotechnology Commercialization Best Practices. Retrieved November 22, 2010, from http://www.quantuminsight.com/papers/030915_commercialization.pdf.

Nanotechnology Database: Sponsored by the National Science Foundation. Retrieved August 2010 from http://www.wtec.org/loyola/nano/dbase.htm.

Nanotechnology Industries. Retrieved June 2010 from http://www.nanoindustries.com/links/government.html.

Nanotechnology Knowledge Gaps. 2006. *Environmental Science and Technology* 40(22): 6871.

Nanotechnology in Massachusetts, A Report on Nano-Scale Research and Development and Its Implications for the Massachusetts Economy. 2004. Massachusetts Technology Collaboration and Nano Science and Technology Institute (NSTI).

Nanotechnology Patents Spreading but Regulatory Framework Needed. Intellectual Property Watch. Retrieved February 2011 from http://www.ip-watch.org.

Nanotechnology Research and Education Centers. Retrieved May 5, 2010, from http://www.nsti.org/outreach/organizations.html.

Nanotechnology Risk Management: Toxicity and Safety. Retrieved January 3, 2011, from http://www.iom-world.org/consulting/toxicity_and_safety.php.Nano-technology in Space Exploration: Report of the National Nanotechnology Initiative Workshop, August 24–26, 2004.

Nanotech Report, An Insiders Guide to the World of Nanotechnology. 2002. *Forbes.*

Nanowerk News: Nanotechnology Metrology and Standardization Institute Opens in Russia. Retrieved March 9, 2010, from http://www.nanowerk.com/news/newsid=7997.php.

National Institute for Occupational Safety and Health (NIOSH). Approaches to Safe Nanotechnology. http://www.cdc.gov/niosh/topics/nanotech/safenano/.

National Nanotechnology Advisory Panel. 2005. *The National Nanotechnology Initiative at Five Years.* President's Council of Advisors on Science and Technology.

National Nanotechnology Initiative (NNI). Retrieved June 15, 2010, from http://www.nano.gov/.

National Nanotechnology Initiative Workshop. 2003. Nanotechnology and the Environment.

National Technology Initiative. 2006. *Environmental, Health, and Safety Research Needs for Engineered Nanoscale Materials.*

National Venture Capital Association. 2005. Fastest Growing Regions for VC Outside of Silicon Valley. Retrieved June 15, 2010, from http://www.nvca.org/index.php?option=com_docman&task=cat_view&gid=100&Itemid=317

New Energy and Industrial Technology Development Organization (NEDO). Retrieved November 5, 2010, from http://www.nedo.go.jp/english/.

Oberdörster, E., et al. 2005. Ecotoxicology of Carbon-Based Engineered Nanoparticles: Effects of Fullerene (C60) on Aquatic Organisms. *Carbon*.

Organization for Economic Cooperation and Development (OECD). Retrieved November 15 from http://www.oecd.org/.

Peters, L., and Sundararajan, M. 2003. Acquisition of Resources for Commercializing Emerging Technologies: Comparing Large Firms with Startups. In *Management of Engineering and Technology, PICMET '03*, pp. 425–435.

Pethig, R. 1996. In *Nanotechnology in Medicine and the Biosciences*, ed. R.R.H. Coombs and D.W. Robinson. London: Gordon and Breach.

Pisanic, T.R., et al. 2007. Nanotoxicity of Iron Oxide Nanoparticle Internalization in Growing Neurons. *Biomaterials*.

Plunkett Research. *Plunkett's Nanotechnology and MEMS Industry*. Retrieved November 15, 2010, from http://www.plunkettresearch.com/Nanotechnology%20MEMS%20materials%20market%20research/industry%20and%20business%20data#trend.

Project on Nanotechnologies. A Nanotechnology Consumer Products Inventory. Woodrow Wilson Center for International Scholars. http://www.nanotechproject.org/index.php?id=44.

Queensland Government, Department of Mines and Energy. Clean Coal Technologies. Retrieved November 2010 from http://www.dme.qld.gov.au/zone_files/Sustainable/cleancoal_6pweb.pdf.

Rashba, E., and Gamota, D. 2004. Anticipatory Standards and the Commercialization of Nanotechnology. *Journal of Nanoparticle Research* 5(3–4): 401–407.

Regis, E. 1995. *Nano: The Emerging Science of Nanotechnology: Remaking the World—Molecule by Molecule*. Little, Brown.

Registration, Evaluation, Authorization and Restriction of Chemical substances (REACH). Retrieved October 22, 2010, from http://ec.europa.eu/environment/chemicals/reach/reach_intro.htm.

Research and Markets. Nanotechnology-Global Strategic Business Report. Retrieved October 11, 2010, from http://www.researchandmarkets.com/.

Responsible NanoCode. Retrieved February 2010 from http://www.responsiblenanocode.org/.

Reuters. 2000. Celera Intentions Misunderstood. *Wired News*, April 6. www.wired.com/news/technology/0,1282,35500,00.html.

Roco, M.C. 2005a. The New Engineering World. *Mechanical Engineering*, April 2005, pp. 7–11.

Roco, M.C. 2005b. Subcommittee on Nanoscience, Engineering and Technology (NSET) and National Science and Technology Council Presentation. Retrieved from Roco, M.C. 2004. The U.S. National Nanotechnology Initiative after 3 Years (2001–2003). *Journal of Nanoparticle Research* 6: 1–10.

Roco, M.C. 2006. Nanotechnology's Future. *Scientific American* 295(2): 39.

The Royal Society. 2004. Nanoscience and Nanotechnologies: Opportunities and Uncertainties. http://www.nanotec.org.uk/finalReport.htm

The Royal Society. Report of a Joint Royal Society—Science Council of Japan Workshop on the Potential Health, Environmental and Societal Impacts of Nanotechnologies. Retrieved January 3, 2011, from http://www.scj.go.jp/ja/int/workshop/summary2005.pdf.

Russia-Kazakhstan Nanotechnology Venture Fund Moves Closer to Founding. Investment, Russian Corporation of Nanotechnologies (RCNT). Retrieved November 2010 from http://.nanotechwire.com/news.asp?nid=10706.

Ryman-Rasmussen, J.P., et al. 2006. Penetration of Intact Skin by Quantum Dots with Diverse Physicochemical Properties. *Toxicological Sciences.*

Schlatka, B. Nanotechnology: Realizing the Promise of Universal Memory. *Nanotechnology Law and Business* 2(3): 1–7.

Serrato, R., Hermann, K., and Douglas, C. 2005. The Nanotech Intellectual Property Landscape 3rd ser. 2.2 (2005). *Nanotechnology Law and Business.* Retrieved March 2011 from http://pubs.nanolabweb.com/nlb.

Service, R.F. 2004. Nanotechnology Grows Up. *Science* 1732.

Skeel, D. Behind the Hedge. Retrieved December 2010 from http://www.legalaffairs. org/issues/November/December-2005/feature_skeel_novdec05.mps.

Smith, A. 2009. *Wealth of Nations.* Complete and unabridged. New York: Classic House Books.

Social Capital Gateway Newsletter. Retrieved March 17, 2010, from http://www.social-capitalgateway.org/.

Small Times Media LLC. 2004. The Rational Investors Guide to Nano and Micro Technology. *Small Times Media,* 18.

Strohal, R., et al. 2005. Nanocrystalline Silver Dressings as an Efficient Anti-MRSA Barrier: A New Solution to an Increasing Problem. *Journal of Hospital Infections.*

Sweeney, R. Bridging Nano Commercialization's Valley of Death. Keystone Edge: Home. Retrieved January 2011 from http://www.keystoneedge.com/features/pananomaterials0218.aspx.

Taylor, M. 2006. *Regulating the Products of Nanotechnology: Does FDA Have the Tools It Needs?* Woodrow Wilson International Center for Scholars Project on Emerging Nanotechnologies.

Toumey, C. 2004. Narratives for Nanotech: Anticipating Public Reactions to Nanotechnology.

Venture Funds Raised 14 Percent Less in 2010. Retrieved February 2011 from http://jetlib.com/news/2011/01/12/venture-funds-raised-14-percent-less-in-2010/#disqus_thread.

Vinge, V.S.S. 1993. The Coming Technological Singularity: How to Survive in the Post-Human Era. Retrieved January 10, 2011, from http://www-rohan.sdsu.edu/faculty/vinge/misc/singularity.html.

Wang, J. 2008. *Resource Spillover from Academia to High Tech Industry: Evidence from New Nanotechnology-Based Firms in the U.S.* Ann Arbor, MI: UMI, Proquest Information and Learning Company

Washizu, M. 1996. In *Nanofabrication and Biosystems,* ed. H.C. Hoch, L.W. Jelinski, and H.G. Craighead. Cambridge: Cambridge University Press.

Weisheng, L., et al. 2006. Toxicity of Cerium Oxide Nanoparticles in Human Lung Cancer Cells. *International Journal of Toxicology.*

Weiss, R. 2005. Nanotech Is Booming Biggest in U.S. *Washington Post,* A6.

Wejnert, J. 2004. Regulatory Mechanisms for Molecular Nanotechnology. *Jurimetrics Journal* 44: 4.

Woodrow Wilson International Center for Scholars Project on Emerging Nanotechnologies. 2006. *Attitudes toward Nanotechnology and Federal Regulatory Agencies.* Peter D. Hart Research Associates.

Index

3D images, 119
12BF Holdings, 121, 122

A

Academic institutions. *See also*
    University laboratories
    patent process services for, 83
Accelerated examination support
    document (AESD) preparation,
    80
Accidents, inevitability of, 146
Accountability, 170
Accountants, encouraging challenges
    from, 142–143
Accounting, 141, 142–143
Accounts receivable, 47
Accrual basis accounting, 47
Acquisitions
    biomedical field, 70
    investment through, 129
Acrongenomics, 30
Acronyms, 182–183
Active shield, 190
Added value, 126
Adhesion, 190
Advertising expense, 50
Advisors, 47
Aerobotics, 190
Aerobots, 190
Agendia, 30
Agreement in Trade-Related Aspects of
    Intellectual Property Rights, 78
Agriculture, nanotechnology products
    in, 25
Agrifood Nanotechnology Research
    and Development Database, 25
Alabama, nanotechnology companies,
    32
Albert Einstein College of Medicine,
    28, 29
ALD NanoSolutions, 32

All-Russian Research Institute
    for Standardization and
    Certification in Mechanical
    Engineering, 114
Alnis BioSciences, 33
Amazon.com, one-click patent, 72
American depository receipts (ADRs),
    136
American National Standards Institute
    Nanotechnology Standards
    Panel (ANSI0NSP), 114, 120
American Society for Testing and
    Materials (ASTM), 116, 118
    nanotechnology standards, 120
Amino acids, 190
Amroy, 30
Angel Capital Association, 125
Angel investors, 124, 135, 147
    conferences, 148–149
    differences from venture capitalists,
        126
    investment criteria, 126
    involvement in business
        management, 125
    as recent trend, 125
    timeline expansions, 127
Ant retina, 4
Antioxidants, 190
Antitrust laws, patent pools and, 74
Arabian swords
    carbon nanotubes in, 19
    nanotechnology in, 18
Arkema, Inc., 38
Arrowhead Research Corporation, 70
Artificial intelligence, 21, 190
Asbestos fibers, similarities to carbon
    nanotubes, 106, 108
Assembler, 190
Atom, 190
Atom clusters, 6
Atomic arrangement, as foundation of
    technology, 92